D0203628

URBAN SEGREGATION AND THE WELFARE STATE

Segregation, social polarisation and social exclusion are central concepts in today's urban debates, dominating discussions on urban transformation and urban realities.

Urban Segregation and the Welfare State examines both developing and existing ethnic and socio-economic segregation patterns, social polarisation and the occurrence of social exclusion in major cities in the western world. Leading contributions from across North America and Europe provide in-depth analysis of particular cities, ranging from Johannesburg, Chicago and Toronto to Amsterdam, Stockholm and Belfast. The authors highlight the social problems in and of cities, indicating differences between nation-states in terms of economic restructuring, migration, welfare state regimes and 'ethnic history'.

Discussing fundamental questions relating to causes, effects and possible future interventions in areas of exclusion and segregation, this book offers a uniquely international perspective on the central debates concerning the social composition of the city, the role of the welfare state and potential policy interventions by the state or local government for transforming the city in the future.

Sako Musterd is Professor of Applied Geography and Planning and **Wim Ostendorf** is Associate Professor of Urban Geography at the Amsterdam Study Centre for the Metropolitan Environment (AME), University of Amsterdam.

URBAN SEGREGATION AND THE WELFARE STATE

Inequality and exclusion in western cities

Edited by
Sako Musterd and Wim Ostendorf

London and New York

First published 1998
by Routledge
11 New Fetter Lane, London EC4P 4EE

Simultaneously published in the USA and Canada
by Routledge
29 West 35th Street, New York, NY 10001

Typeset in Galliard by
Florencetype Ltd, Stoodleigh, Devon
Printed and bound in Great Britain by
TJ International Ltd, Padstow, Cornwall

British Library Cataloguing in Publication Data
A catalogue record for this book is available from the British Library

Library of Congress Cataloguing in Publication Data
A catalogue record for this book has been requested

ISBN 0–415–17059–1

CONTENTS

FIGURES

TABLES

CONTRIBUTORS

Frederick W. Boal is Professor at the Department of Geography of the Queen's University of Belfast

Lars-Erik Borgegård, **Eva Andersson** and **Susanne Hjort** are Associate Professor and research assistants respectively, at Uppsala University

Anthony J. Christopher is Professor at the Department of Geography of the University of Port Elizabeth

Susan S. Fainstein is Professor at the Department of Urban Planning and Policy Development of Rutgers – the State University of New Jersey

Jürgen Friedrichs is Professor of Sociology at the University of Cologne

Chris Hamnett is Professor of Urban Geography at the Department of Geography of King's College London

Jerome L. Kaufman is Professor at the Department of Urban and Regional Planning of the University of Wisconsin – Madison

Christian Kesteloot is Professor at the Institute of Social and Economic Geography of the Catholic University of Leuven

Robert A. Murdie is Associate Professor at the Department of Geography of York University, Toronto

Alan Murie is Professor at the School of Public Policy and Director of the Centre for Urban and Regional Studies (CURS) at the University of Birmingham

Sako Musterd is Professor of Applied Geography and Planning at the Amsterdam Study Centre for the Metropolitan Environment (AME) of the University of Amsterdam

Wim Ostendorf is Associate Professor of Urban Geography at the Amsterdam Study Centre for the Metropolitan Environment (AME) of the University of Amsterdam

Paul White is Professor of Geography at the Department of Geography of the University of Sheffield

Herman van der Wusten is Professor of Political Geography at the Amsterdam Study Centre for the Metropolitan Environment (AME) of the University of Amsterdam

PREFACE

The study of urban segregation and social exclusion issues is 'core business' to urban geographers, urban sociologists, housing specialists and urban economists. If some emphasis is given to the possible role of the state or local governments in transforming the city and its social composition, then political scientists are also warmly welcomed to complete the list. We had that wide variety in mind when, a few years ago, we – the Amsterdam Study Centre for the Metropolitan Environment (AME) of the University of Amsterdam – organised an international seminar on the issue of 'Segregation and the welfare state; inequality and exclusion in western cities'. We also had in mind that the seminar should be truly international. This not only offered the opportunity to learn from situations which are different from the one each of us is most familiar with, it also offered the opportunity to get some grip upon the relative importance of the role of welfare states. From the start of the seminar, that was what we wanted to address in particular.

The authors were invited to discuss fundamental questions about 'causes' and 'effects' and possible interventions in the fields of exclusion and segregation, with special reference to the role of the welfare state. We attached a fairly strict outline to the invitation, culminating in a list of questions the authors were asked to address. We would like to thank them, since they adhered to the outline so well, and have made stimulating contributions to the debate about segregation and the welfare state. In our opinion they succeeded in producing a most homogeneous book. Of course this also expresses their joint interests.

The British authors deserve extra thanks, since they were so kind as to be not only critical, as usual, but also helpful in correcting the English of texts of non-native English-speaking authors. We regard their support as an indication of the strength of the network that developed with the seminar.

We hope the book will serve to increase the understanding of the role of the welfare state in transforming cities.

Amsterdam, June 1997
Sako Musterd & Wim Ostendorf

ACKNOWLEDGEMENTS

Neither the seminar nor the book could have been realised without the support of several people and institutions. Organisational support was provided by Saskia Pelletier and Lotty Jansen Saan as well as by the producers of the graphics: Hans de Visser and Christian Smid. We owe special thanks to Dick van der Vaart of the AME, who not only organised all the practical aspects of the seminar, but also was extremely helpful in putting the texts into the right format, and in pushing the authors if that appeared to be necessary. Financial support was gratefully accepted from the municipality of Amsterdam, which sponsored the seminar within the framework of the Honorary Wibaut Chair at the University of Amsterdam.

We would also like to thank Walter De Gruyter and the Canada Mortgage and Housing Corporation for their permission to reproduce the illustrations presented in Figures 5.1 and 5.2.

1

SEGREGATION, POLARISATION AND SOCIAL EXCLUSION IN METROPOLITAN AREAS

Sako Musterd and Wim Ostendorf

Urban social consequences of the restructuring of society

Segregation, social polarisation and social exclusion are central concepts in today's urban debates (Wilson 1987, Sassen 1991, Fainstein, Gordon and Harloe 1992, Massey and Denton 1993, Hamnett 1994a, Marcuse 1996, O'Loughlin and Friedrichs 1996). In many countries these concepts have not only dominated the urban transformation debate for a long time but, according to many people, urban realities too, and they still do. Cities in the western world in one way or another reflect the socio-spatial outcomes of polarisation, segregation and exclusion processes. The outcomes vary according to the character and intensity of the social processes. In their turn these social processes depend upon a wider range of factors and developments. The economic structure of a city and the kind of restructuring that is going on are frequently seen to be among the most powerful forces behind social fragmentation and integration in the urban realm. However, welfare state regimes and the current changes (cutbacks) noticeable in these areas are also thought to be very important. Other related factors are frequently mentioned too, such as the racial or ethnic population division in society, and the reinforcing effect of socio-spatial and ethnic segregation itself.

In this book the attention will be explicitly focused upon the relationship between segregation, polarisation and exclusion on the one hand and the structure and transformation of the welfare states on the other. However, welfare state regimes and economic structures as well as these other factors are all interrelated. Therefore, it is difficult and possibly unwise to try to isolate just one of these factors. Focusing upon welfare states is merely a matter of emphasis rather than an attempt to restrict the debate. This focus may be regarded as an effort to counterbalance what we regard as the over-attention on social polarisation and segregation as a result of economic restructuring and globalisation processes.

In this chapter we will briefly introduce the major issues dealt with in this book, starting with the central concepts set out in the next section. In that section we will refer to the dimensions which are said to be key issues in the theoretical debates about the forces underlying urban social processes. We also discuss the issue of the 'myth or reality' of social polarisation and exclusion. We will touch upon that in another section. In the final section we will briefly introduce each of the case studies which form the body of this book in chapters 2–13.

In chapter 14, the final chapter of this book, some conclusions are drawn by Herman van der Wusten, who chaired the conference which gave birth to this book, and Sako Musterd, the initiator of the conference.

Concepts and theoretical dimensions

Economic restructuring

Over the past decade, the advanced industrial countries of the world have all gone through a process of economic restructuring that is frequently assumed to be strongly associated with the process of globalisation. Improved technological conditions have led to a growing interconnectedness and internationalisation of firms and economic processes, which are expressed in the rapid growth of flows of people, money and goods across the world. Among the characteristics of these changes are a growth in the demand for services and thus for service jobs, for which highly skilled labour is required. However, the global economic restructuring process frequently also brings unemployment as well as a demand for low-skilled or unskilled jobs. The final result of the restructuring process is said to be increased *social polarisation*, that is, a growth in both the bottom end and the top end of the socio-economic distribution, for example an increase in the proportion of households with low skills or low income (many of whom are immigrants) and at the same time an increase in the proportion of people who are highly skilled or the number of households with high incomes (e.g. Sassen 1991). Increased social inequality and social division results in the social inclusion of one part of society and the *social exclusion* of another part. The excluded lose the opportunities, the means and finally the ability to participate in society, which is expressed by a lack of labour market participation, low school participation, a weak position in the housing market, limited political participation and restricted socio-cultural integration. The divisions are also said to be reflected in spatial patterns. Separate residential concentrations of wealthy people and of poorer households will result in socio-economic *segregation*.

Race, ethnicity and immigrants

However, the picture is much more complicated since in many places, apart from class divisions, there are also significant racial, ethnic and/or immigrant divisions and these are also expressed spatially in segregation patterns. The most extreme expressions of racial, ethnic or immigrant divisions are without any doubt the Black ghettos in the USA and South Africa. Less extreme, but still significant spatial concentrations of immigrants are encountered in Europe and also in the USA.

Although there appears to be a strong association between the racial and immigrant divisions on the one hand and the socio-economic divisions on the other – Blacks and immigrants, for example, often have a relatively weaker employment position compared to other population groups – that association is not universal. It is expressed by the segregation patterns of class and race (or ethnicity). Denton and Massey (1988) and Morrill (1995), for example, have shown that high levels of ethnic segregation persist for each income class that is distinguished, especially for Blacks. Earning a higher income had only slight desegregation effects for Hispanics and Asian people.

An interesting feature, however, is that the position of immigrant groups coming from developing countries relative to the position of the indigenous population, which is also spatially expressed in segregation patterns, appears to be different between different states. The levels of inequality and the levels of spatial segregation seem to be much lower in European countries, compared to the North American situation. Consequently, Peach (1996), Rhein (1996) and Wacquant (1993), for example, notice that, although many Europeans are concerned about the development of ghettos in European cities, the actual situation is still far from that reality, possibly because of the specific European context. However, states within Europe are by no means identical (Esping-Andersen 1990). There are differences in terms of the organisations and histories of the welfare states in all parts of the world.

The welfare state

After the Second World War extensive systems were developed in many European countries aimed at reducing the social risks of illness, disability and unemployment. The early initiatives of Bismarck in the field of financial assistance in case of illness, as well as the social security systems that were developed as the basis of Beveridge's proposals in Britain in 1941, are frequently mentioned. After the initial efforts in welfare provision, some countries developed very extensive welfare systems. Together with these developments a substantial redistribution of bargaining power was established. Income taxes that were both progressive and high, the development of high minimum wage levels, the provision of relatively generous benefits in old age, illness, unemployment and disability, extensive systems to redistribute the costs and the benefits in the sphere of housing (brick and mortar subsidies, individual rent subsidies) were all made part of the system of state involvement in many countries. Most Western European states experienced tremendous welfare expansion between 1945 and the mid-1970s.

It is often argued that there is a strong relation between the extent to which the welfare states have developed their social security and welfare systems and the levels of social polarisation, socio-spatial segregation and social exclusion in urban areas. In general, there is a belief that well-developed welfare states have thus far been successful in shielding certain population categories from social deprivation and isolation.

However, it is not only the structure of the welfare state we should look at, but also the changes which occur over time. After the first (1973) and particularly after the second (1979) oil crisis, many western countries, in particular those with relatively well-developed welfare provisions, were faced with relative economic decline and economic restructuring, and rising economic problems. Higher structural unemployment in the large cities, particularly among immigrants, and the growth of state budget deficits, which together seem to be the result of these processes, laid the basis for the revision of many welfare states. During the past fifteen years many nation-states which had developed welfare provisions slowly started to move in a more neo-liberal direction. A general atmosphere was created in which many initiatives were pushed into more market-led and deregulated directions. Tax reduction, no universal welfare benefits, a reduction of redistribution, deregulation, subsidy cuts and more flexible labour markets became the new keywords. It is now a widely shared view that these revisions will result in an increase in individual employment opportunities, but will at the same time result in an increase in social polarisation and socio-spatial segregation in urban areas. In an issue of *Built Environment* (1994), entitled 'A rising European underclass?', all contributors not only noticed a tendency to liberalisation in the German, British, Swedish, French, Belgian and Dutch welfare states, but also pointed to the relationship between the development of ethnic and socio-economic spatial segregation in metropolitan areas and the development of urban social problems. However, firm empirical support for the relationship could not be shown.

One of the crucial elements in today's theoretical debate about urban social problems is the question of the relation between global economic restructuring processes and the role of the (welfare) state. A key hypothesis in the literature is that the globalisation process almost inevitably results in an increase in the power of 'the market' (private firms) and a loss of power and opportunities of local and national governments. However, several alternative hypotheses may be formulated with regard to the role of the state. Peter Marcuse (1996: 40) suggested that

> it is not, however, a reduction in the role of the state; to the contrary, it may even be an increase in that role. . . . They rather shift direction, from a social and redistributive to an economic and growth or profit-

> supporting purpose. At the same time, they shift from a public, in the sense of democratic or popular, instrument, to an instrument of private business purpose.

Such a viewpoint illustrates that economic factors are not necessarily the dominant forces *per se*. One might even go one step beyond that and think of state intervention as a much more independent factor, with effects upon the social structures possibly irrespective of economic forces.

The reinforcing effect of spatial segregation

Apart from the mutual relations that are expected to exist between economic restructuring, ethnic or racial divisions and the changing nature of welfare states on the one hand, and social polarisation, socio-spatial segregation and social exclusion on the other, it is also assumed that sharp socio-spatial or ethnic segregation of the population is a potential major cause of social problems for individual households in cities. However, in this respect much thinking seems to be inspired by images and perceptions that originate from specific, but relatively extreme cases, i.e. specific cities in the United States of America, where social polarisation and social and ethnic segregation reach very high levels. In those circumstances segregation may easily become a factor in its own right. Massey and Denton (1993: 9), in their book *American Apartheid*, even argue that 'Residential segregation is the principal organisational feature of American society that is responsible for the creation of the urban underclass'. In their view, segregation creates the structural conditions for the development of a kind of counterculture in which a job, good education and strong relations between people are no longer part of the prevailing system of values and norms. Schill (1994: 443), too, expressed this view: 'this concentration of poverty generates attitudes, behaviors and values that impede the ability of residents to grasp whatever opportunities exist for social mobility.' Geographically oriented sociologists, too, believe that segregation in American cities is producing and aggravating social problems (Wilson 1987).

However, as we have said, we have to remember that the ethnic and socio-economic spatial segregation of the population in US cities is generally more rigid compared to the segregation encountered in many other western cities, particularly those in Europe. If it is true that the so-called redistribution welfare states of continental Europe, which are also different in social, political and ethno-cultural terms, have produced cities that are only moderately segregated in the first place, it is questionable whether these moderately segregated areas have any effect on processes of social integration or exclusion at all. Western countries should not be treated as if they were all the same. Of course, this holds true for the states within Europe too.

It is within this broad framework that the aim of the book has been formulated: *to demonstrate and discuss the factors that are relevant to the process of socio-spatial segregation, social polarisation and social exclusion of parts of urban society, with special attention to the effects of different and changing welfare states.*

Myth and reality

The reader who is familiar with the subject dealt with in this book will have noticed that over the last decade, and certainly over the past few years, there has been renewed concentration on the problems of social polarisation and social exclusion. Again and again, newspapers, magazines, books and scientific journals discuss the growth of urban social problems. Although that attention is clearly related to an increase in serious social problems in the cities, there is some bias in it as well. Concepts such as socio-spatial segregation, urban underclass, social exclusion, social polarisation, deprivation and marginalisation have become fashionable rhetoric, and can therefore serve many purposes. We should not forget that many comments and texts are hardly more than an expression of fear and a reflection of the predominant discourse, the hegemonic way of thinking at the time, that may well cover a lack of understanding. A good example of the latter was provided by the *Economist* a few years ago, when it paid attention to 'Europe and the underclass' (July 1994). It concluded, among other things, that

> As yet, Western Europe does not have an urban underclass to compare with that of the United States. But the growth of long-term unemployment seems to be dragging it inexorably in that direction. In cities across Western Europe – such as Frankfurt, Berlin, Lyon, Paris, Amsterdam and Liverpool – the shadowed lives of the urban poor are getting darker.

Although it was admitted that United States cities are still the prime examples of social problems and underclass, the journalists advised, as a solution to prevent European cities from following the American examples, adopting the American version of the economic and social welfare state!

> Europe has priced much of its labour force out of employment, compensating it with welfare payments. Only a thoroughgoing reversal of that strategy can do much to get Europe's unemployed off the park bench and back into work. Encouraging the kind of dynamic economy in which lots of jobs are created will mean hacking away at policies that have long operated in favour of rigid work rules, high social costs, subsidies and protectionism.
>
> (*Economist*, 30 July 1994)

However, first underlining the unsatisfactory situation in United States cities, then advising the adoption of the American model, does raise some questions.

Examples of the mixing of myth and reality abound. In the Netherlands scientists and politicians have expressed their fear of increased ethnic and socio-economic segregation for some decades now. Dutch research projects, however, have repeatedly shown a high degree of stability in the levels of (ethnic and socio-economic) segregation. Sometimes the spatial picture appeared to be changing, but the overall level of segregation has changed little (van Amersfoort 1992, Musterd and Ostendorf 1994). Another example that indicates that one should be cautious with broad observations in this field is provided by empirical analyses of the process of social change in London. According to Hamnett (1994a and chapter 2 of this book), increased occupational polarisation could not be shown.

Of course, these findings are also a function of the way the concepts are measured and defined. Many specific issues are related to the operationalisation of such concepts. Are persons addressed or households? Is social position best indicated by income, education, employment or profession? What kind of spatial unit is involved? In all the contributions in this book attention is given to these kinds of problems of defining concepts. Yet in two chapters conceptual problems get pride of place. Hamnett (in chapter 2) highlights the variable meanings of the concept of social polarisation in different contexts, whereas White (in chapter 9) addresses ideological and conceptual issues related to social exclusion.

Case studies on segregation and social exclusion

In this volume the factors mentioned above provide the basic elements for the discussion of social problems in and of the city. Attention will be paid to existing and developing ethnic and socio-economic segregation patterns in major metropolitan areas in the western world, to the description of the development of social polarisation, and to the occurrence of social exclusion. Among the dimensions supposedly crucial for understanding segregation, polarisation and exclusion, types of welfare state and state involvement differentiate most clearly between the cases dealt with in this volume. In order to emphasise that dimension, the chapters are loosely structured on the basis of the type of welfare state involved. If a – crude – continuum is considered to exist, ranging from the most liberal types of state via the conservative corporatist types to the more social democratic types of state, the countries of cities analysed in this volume can be ranked as follows: USA, Canada, UK, Belgium, France, Germany, the Netherlands and Sweden. The first states are associated more clearly with free market enterprise and few state initiatives to reduce segregation and inequality. The latter states represent much greater state involvement in a wide variety of areas, which may have helped to reduce segregation and inequality. Of course, if these states are looked at in more detail, each of them has its own distinctive profile. Reality proves to be much more complicated than this simple continuum suggests. However, given the idea that the type of state involvement is of great importance to the understanding of segregation and exclusion, it makes sense to present this order as a structuring mechanism. However, we will start the series of case studies with the contribution of Hamnett, illustrated with material on London. He not only very clearly sets out the 'conventional wisdom' of the relation between social polarisation and global economic restructuring, as set out in Sassen's work, but also convincingly argues for a new perspective accounting for differences in national welfare provision.

The South African case has deliberately not been mentioned so far. This case could be dealt with in the same way as the other representatives of cities subject to liberal state regimes, since in economic terms South Africa may also be labelled 'market-oriented' and in that sense 'liberal'. However, South Africa has also shown very strong state intervention in the field of population distributions and in that sense is not comparable with other contemporary liberal states. Segregation was deliberately developed and supported by government law and highly institutionalised. Consequently, South African cities are clearly the most extreme expressions of ethnic and socio-economic segregation and of social and other forms of exclusion. This situation frequently resulted in inter-group tensions, which have in turn been used as a sign to sharpen segregation. The reason for giving this case a separate position is therefore clear. Segregation was meant to develop and was meant to exclude, meant to protect the values of the dominant population category against what was defined as the inferior category. The current transformations, however, reveal the power of the resistance to the development of such separated societies.

As is also illustrated in the other cases dealt with in this volume, segregation and exclusion processes are partly the result of the way in which the state responds to what happens in society. In all situations, except for the South African case, state involvement is directed by an encompassing democratic framework. One might argue that socially driven mechanisms are the dominant forces there. These mechanisms include cultural (ethnic) separation processes as well as the working of state institutions and socio-economic – class – differentiation processes, and discriminatory elements are also relevant. But segregation is not covered by accompanying 'ethnic' law. A bottom-up process has resulted in some form of ethnic and socio-spatial segregation and exclusion instead.

The core of this book is formed by twelve case studies (chapters 2–13), which together cover the western states on both sides of the Atlantic: North America, Western Europe and South Africa. Each of the cases deals with one or more metropolitan areas in these three parts of the world. The metropolitan areas include Chicago, New York and several other cities in the United States; Toronto in Canada; Belfast, Edinburgh and London in the United Kingdom; Brussels in Belgium; Paris in France; Hamburg in Germany; Amsterdam in the Netherlands; Stockholm in Sweden; and Port Elizabeth and several other cities in South Africa.

Crucial elements of the contributions are synthesised in the final chapter, 14. There an attempt is made to compare the cases dealt with by referring to the structure and transformation of the welfare state models involved, within the context of differences between nation-states and between cities within nation-states in terms of economic restructuring and position in the international migration processes and 'ethnic history'. In the remaining part of this section each of the case studies involved will be introduced briefly.

The issue of myth and reality, previously referred to, is clearly addressed in the contribution by Chris Hamnett (chapter 2). In a critical analysis of the use of central concepts in social polarisation theory and exclusion debates, he confronts the reader with empirical evidence of the lack of care with which basic concepts have been used. In his contribution he provides data on London, but also paves the way to a broader comparison of continental European countries with the United States. Furthermore, he stresses the importance of investigating the differences between different welfare state regimes.

His focus is on social polarisation rather than on social segregation *per se*. It is argued that the forms of social polarisation in different countries are not homogeneous or unidirectional. Special attention is given to the distinctions between occupational and income structures, and to the division between the economically active and inactive and the unemployed. It is argued that income polarisation and occupational polarisation are not related to each other in a one-to-one way. Income polarisation may, for example, also be the result of changes in taxation, welfare benefits, unemployment or shifts in household composition and age structure of the population.

Susan Fainstein, in chapter 3, argues that the history of group inclusion and peripheralisation – of Blacks in particular – is one of the most dominant factors influencing the development of segregation in United States cities. It is stressed that the United States has a double-sided history. The first side is famous for its assimilation of voluntary immigrant foreigners and is in that respect also quite different from many European states, which have developed much stronger national cultures and are much less open to assimilation of foreigners; the second side is infamous for its slave-holding period in which Blacks were forced to involuntary servitude to White rulers. And it is this double-sided history that is expressed in the ethnic and social structure of the United States and its cities and in the bad prospects for the concentrations of poor Blacks in particular.

Large concentrations of Blacks also developed elsewhere, for example in South Africa, discussed in chapter 13 of this book. However, there are major differences between the United States and South Africa. Apart from the specific attitudes against immigrants and Blacks, another major point of difference is the absence of any specifically targeted law that has stimulated ethnic segregation and separation processes in the United States, at least during the twentieth century. The absence of such laws did not, however, result in a reduction in the level of social and ethnic segregation or in the level of social exclusion of Blacks. Obviously, a different set of welfare provisions was developed in the United States compared to other western countries.

A somewhat extreme case of 'hyper' segregation in the United States, but also a good example of the American metropolises, is the segregation in the metropolitan area of Chicago, which Jerry Kaufman deals with in chapter 4. In Chicago, the segregation of African-Americans and the effects of that segregation are of a clearly different character from that of Hispanics or other (former) immigrants. Kaufman mentions several reasons why segregation of Blacks is so evident in Chicago. Among them are racism and economic restructuring, but also the fragmented governmental system (the competition for tax income) and governmental actions (renewal – or destruction – of (viable) low-income neighbourhoods) are mentioned, as are the redlining actions of banks and insurance companies, restricting financial flows to Black low-income areas and the creation of massive public housing developments in low-income areas.

The effects of the 'hyperghettoisation' are discussed on the basis of a review of in-depth research done by Wilson and others and by Massey and Denton. The effects are expressed in several indicators of deprivation and livability and assumed to be rooted in joblessness, in a decline in the social organisation of the areas involved and, in Massey and Denton's work in particular, in racial segregation itself. The effects of an increase in poverty among poor Blacks in different racially segregated contexts are shown: spatial concentration intensifies the problems. The economic restructuring (lower income, higher poverty rates, more joblessness for Blacks) interacts with segregation.

Bob Murdie focuses attention on Toronto, Canada's major financial and economic centre and the country's major reception area for immigrants (chapter 5). He outlines three important differences between Canada and the United States, which are reflected in the spatial segregation and social exclusion of minority groups in Toronto. One is the wider development of social welfare programmes in Canada. A second is the less dominant role that race has played in the development of Canadian cities. And finally, Toronto is not characterised by the American level of political fragmentation.

As in the United States, the initial immigrant flows were from Europe, but now immigrants from Asia, Africa, South and Central America and the Caribbean dominate. These recent immigrants exhibit high levels of spatial segregation although economically they are a very diverse group. They are located in the traditional immigrant reception area near the downtown core and (a majority) in a variety of suburban areas. Although European groups such as the Italians and Portuguese remain segregated residentially, they have achieved comparatively high levels of spatial and social mobility. Overall, in Toronto, there is no evidence that spatial segregation by itself is a deterrent to social mobility.

Recently arrived refugees such as the Somalis are in greatest danger of social exclusion. They exhibit multiple disadvantages (low incomes, high unemployment, racial discrimination) and also had the misfortune of arriving during a severe downturn in the economy. More and more marginalised groups tend to live in public housing. Because of the relatively small amount of public housing in Toronto, the chance of this group becoming socially residualised is high.

Belfast is known as the UK's, but also Europe's, most segregated city. As Frederick Boal underlines in chapter 6, the sharp segregation is expressed first and foremost in ethnic (Catholic and Protestant) terms, but also appears to exist in class terms. The background to these expressions of segregation is quite distinctive. The segregation of Catholics and Protestants, which has increased sharply in recent years, is strongly related to responses to clashes between nationalities and subsequent outbursts of violence. The class segregation within each of the ethnic groups seems to be more comparable to experiences in United States cities. Combination of the two types of segregation reveals that the two most severely segregated segments of Belfast's population are the low-income Catholics and the low-income Protestants. Boal notes that they are class-segregated from their own middle classes and they are ethnically segregated from each other. The British welfare state has not been able to put ethnic strife aside.

Boal also pays attention to some of the positive sides of segregation. He refers to Wacquant and Wilson's distinction between the 'organised ghetto' and the 'hyperghetto'. The latter is the expression of the underclass, from which practically no one is able to escape. The first type of ghetto, however, has a more appealing connotation. An organised ghetto may help to provide feelings of security, to maintain other lifestyles, to support ethnic entrepreneurship, or to provide an organisational basis and a critical mass for action in a wider society and for the development of vital subcultures.

The role played by changing housing provision in stimulating and preventing segregation and social exclusion is dealt with extensively by Alan Murie (chapter 7). The author rightly stresses housing as an important dimension of welfare state arrangement. The contribution starts with a critical examination of the most influential welfare state typology of Esping-Andersen. In particular, Murie notes that welfare arrangements in the sphere of taxation systems, health care systems and housing provision and the functioning of the housing market are not represented in that classification, though the interaction with social security arrangements is considerable. Britain would clearly not fit into the class of liberal welfare states, if one looks at the socialised health and the decommodified housing system. He also criticises the lack of attention to the changes in the welfare systems, and again refers to Britain's active (re)commodification in housing, health, pensions and other benefits (selling the welfare state). Increased segregation and social exclusion of part of the population are expected to be associated with these (re)commodification processes. His argument is illustrated with information regarding housing change in Edinburgh, particularly in the council housing areas, and links to changes in the social composition of these areas.

Chris Kesteloot focuses attention upon polarisation and segregation in the Brussels urban region (chapter 8). He puts the discussion in the context of changes in the Belgian welfare state, which became very well developed in the mid-1970s, especially in terms of minimum income provisions. It is shown that social polarisation in Brussels is increasing and is spatially translated into a deepening segregation between social and ethnic groups. Moreover, this is true for each of the three different spatial scales (city–periphery; differentiation between sections of the city; and individual neighbourhoods) at which the analyses were done. Kesteloot shows the city is poorer than the suburbs; the inner-city crescent is poorer than the rest of the city; and some neighbourhoods in this crescent are poorer than other neighbourhoods. In addition, there are major differences between deprived neighbourhoods, which are related to differences in the causes of the problems. Recognition of these differences is not unimportant in the search for local development strategies that are aimed at improving economic conditions, which Kesteloot discusses.

A thorough discussion of the concept of social exclusion can be found in chapter 9, where Paul White illustrates his arguments with examples from Paris. He relates the discourse of hegemonic structures, power and ideologies within society to the concept of social exclusion. He supports an approach in which the distinction between those who have access to resources such as jobs, welfare services, education, housing, territory, citizenship, etc., and those who are excluded from such resources is perceived in terms of power relations and ideologies. He elaborates on national ideologies with regard to immigrants, young people and unemployment; on local Parisian ideologies that are focused on the embourgeoisement of the city; and on suburban ideologies that are dominated by the *grands ensembles*, the large social housing estates, which increasingly face 'territorial stigmatisation'. The exclusionary consequences of the mechanisms that are part of the ideologies are set out and partly illustrated with some data on population change and housing in Paris and its suburbs.

Referring to the macro–micro model of segregation of Coleman, Jürgen Friedrichs, in chapter 10, deals with the question of what factors condition the segregation process. He concludes that attention should be paid to income inequality, inequality in terms of education (and lifestyle) and to discrimination, within the broader framework of economic restructuring and social differentiation. The main elements of the macro–micro model provide the guidelines for the analysis of the Hamburg case, where clear spatial patterns in terms of income and ethnicity are shown. The association between the two distributions is described, as is the association with unemployment. It is clear that even in the most developed welfare states people cannot be shielded from unemployment. Friedrichs also elaborates on the reaction of the population to the social and ethnic transformation processes going on in the city. That reaction was measured by analyses of voting behaviour. Level of education appeared to be the variable most closely related to (extreme) voting behaviour. The relation between voting behaviour and income or ethnicity was less straightforward.

Musterd and Ostendorf (chapter 11) test the hypothesis that in the Dutch situation social participation would not be much affected by the differences between neighbourhoods in terms of the concentration of deprived people. The relation between segregation and social exclusion was analysed, while controlling for other relevant dimensions that were measured at the individual level, such as income, housing, labour market position, education, ethnicity and demography. The empirical analysis focused on social segregation instead of ethnic segregation.

Unexpectedly, the central hypothesis had to be rejected. The effect of the segregation of poverty on social participation appeared to be important. Only small differences in the concentration of poverty and only moderate levels of segregation are sufficient to generate significant differences in the percentage of people who face reduced social participation.

The authors refer to two interpretive theories to explain the results. The first is the theory of stigmatisation of a neighbourhood, through which people are excluded by certain employers, for instance. The second is the type of theory in which the effects of negative role models are predominant. Once again, welfare states are apparently unable completely to suppress such effects.

Sweden, without any doubt once the most advanced welfare state, now seems to be experiencing a phase of drastic decline. Unemployment is rising fast and state deficits are reaching all-time highs. It is within that context that Lars-Erik Borgegård, Eva Andersson and Susanne Hjort deal with socio-economic changes in metropolitan Stockholm in chapter 12. Although it is too early to evaluate the effects of the revolutionary changes within Sweden – the influence of the state on social processes is still large and is still supposed to reduce the effects of economic restructuring – empirical analyses of polarisation and segregation already show significant changes. Data are analysed at several scales. Analyses of income differences at the municipal level show a decrease over time. Municipalities appear to converge in this respect. But if the analysis is carried out at the detailed level of the neighbourhood or housing area, increasing income gaps could be shown. The authors conclude that, at that level, polarisation and spatial segregation have increased during the 1990s.

South African cities, dealt with by Anthony Christopher in the final case study (chapter 13), clearly show a two-layer segregation pattern. The more dominant and infamous layer is formed by ethnic separation, the other by class differences. Segregation indices are among the highest in the world. Ethnic apartheid in particular has developed in an extreme form. It was entirely institutionalised by the dominant political power (cf. the Group Areas Act) and therefore prevented social interaction between different groups from occurring, in a very rigid and legally supported way in all fields of life. Christopher shows the development of that segregation in several South African metropolitan areas as well as the similarly excluding effects of society. Special attention is paid to the case of Port Elizabeth. The deeply rooted apartheid structure imposed by the state has paradoxically become even more visible after the 'voting rights revolution', which took place in 1991. It seems as if desegregation and reintegration will be slow, and will take many decades instead of years. Pessimists even doubt whether any long-term success will be achieved. However, decades of rigid legislated segregation on a racial basis cannot be wiped out overnight. One of the major problems facing South Africa is the current huge differences between Whites and others in terms of social class. On many occasions, class differences have taken over the role of ethnic differences, with almost identical segregation effects.

2

SOCIAL POLARISATION, ECONOMIC RESTRUCTURING AND WELFARE STATE REGIMES

Chris Hamnett

Introduction

In this chapter I want to examine the relationships between urban social polarisation, economic restructuring and the role of the welfare state. The existence of polarisation or dualisation – the growing division in society between the haves and the have-nots; the socially included and the excluded; and a shrinking of the size of the middle groups – has become almost a conventional wisdom regarding social change and divisions in western cities. I want to problematise the notion of polarisation, which is accepted uncritically. As Fainstein *et al.* pointed out:

> The images of a dual or polarised city are seductive, they promise to encapsulate the outcome of a wide variety of complex processes in a single, neat and easily comprehensible phrase. Yet the hard evidence for such a sweeping and general conclusion regarding the outcome of economic restructuring and urban change is, at best, patchy and ambiguous. If the concept of 'dual' or 'polarising' city is of any real utility, it can serve only as a hypothesis, the prelude to empirical analysis, rather than as a conclusion which takes the existence of confirmatory evidence for granted.
>
> (1992: 13)

Second, and following from this, I wish to dispute the way in which the concept of social polarisation and the associated concept of the dual city is commonly used as a general, all-purpose, signifier of growing inequality and social divisions. While I accept that social polarisation has been used in a number of different ways (Pinch 1993, Pahl 1988) in Britain and North America, and that it has an important representational, ideological and rhetorical role regarding growing social divisions in cities, there is a parallel danger that, by uncritically accepting the existence of social polarisation as some sort of general, catch-all, process, we may fail to see the existence of different forms of polarisation in different cities. Social polarisation is not a single, homogeneous process which operates in the same way in different places.

Third, and perhaps most importantly, there is the danger that, by uncritically accepting the conventional wisdom, we may fail to see that the processes driving polarisation in different cities differ/are mediated in various ways. It is thus necessary, in my view, to conceptually unpack the term 'polarisation' and to examine the extent to which different forms of polarisation are found in different contexts and to theorise the reasons for such variations. Otherwise we risk becoming slaves to unexamined, imprecise or ill-defined concepts.

I take 'polarisation' to be a term referring to a change in certain social distributions such that there is a shift away from a statistically normal or egg-shaped distribution towards a distribution where the bottom and top ends of the distribution are growing, relatively and possibly absolutely, at the expense of the middle. This is the dominant interpretation and reflects the concerns of a number of commentators (Harrison and Bluestone 1988, Kuttner 1983, Lawrence 1984, Levy 1987). Marcuse put it well:

> The best image . . . is perhaps that of the egg and the hour glass: the population of the city is normally distributed like an egg, widest in the middle and tapering off at both ends; when it becomes polarised the middle is squeezed and the ends expand till it looks like an hour glass. The middle of the egg may be defined as intermediate social strata. . . . Or if the polarisation is between rich and poor, the middle of the egg refers to the middle income group. . . . The metaphor is not structural dividing lines, but of a continuum, whose distribution is becoming increasingly bi-modal.
>
> (1989: 699)

Mollenkopf and Castells also point out that the dual city notion

> usefully emphasises one trend – both the upper and the lower strata of a given society grow at disproportionate rates. Thus, in the context of the polarisation thesis, the dual city becomes a simple [*sic*] matter of empirical testing of two basic questions:
>
> a. Are the top and bottom of the social scale in a given city growing faster than the middle (with the key methodological issue being how to construct a scale to measure social distribution)?
> b. How does such polarisation, if it exists, translate into spatial distribution at the top and bottom of local society, and how does such specific residential location affect overall socio-spatial dynamics?
>
> (1991: 407)

My concern in this chapter is with the social dimensions of polarisation rather than with social segregation *per se*, but the two questions are clearly linked together. Finally, and following on from the concerns outlined above, I wish to challenge the dominant theory of polarisation advanced by Sassen and others which sees occupational and income polarisation as the outcome of a general shift from manufacturing to services which is particularly marked in global cities as a result of their concentration of key advanced business and financial services, gentrification and the growth of sweated manufacturing. This thesis of polarisation is essentially unicausal, focusing on change in the paid labour force. I shall argue that, whilst there is strong evidence for income polarisation in capitalist economies, there is no evidence of occupational polarisation and that the development of income polarisation may also be the result of changes in taxation, welfare benefits and unemployment rather than occupational restructuring.

The thesis I wish to put forward is that the extent and the forms of social polarisation in different countries are unlikely to be homogeneous or unidirectional; that they result from a combination of economic restructuring which is changing the structure of the labour market, the structure of occupations and incomes in the paid labour market and the division between the economically active and the inactive and unemployed. In addition, I want to argue that there may be other forces generating income polarisation such as shifts in household composition and the age structure of the population. Finally, I want to argue that the extent of polarisation in the occupational and income structure of western societies is likely to be mediated by structures of welfare provision and taxation which are instrumental in influencing both the necessity to enter the paid labour market and the incomes derived from paid employment and from welfare benefits. I further wish to argue that the extent of polarisation in many US cities is a result of the specific institutional context in that country, particularly the high and growing level of immigration and its implication for labour supply, the relative paucity of welfare provisions and income support for the poor, and the absence of effective minimum wage legislation and the growth of a large, low-paid casualised service sector. The rapid growth of income polarisation in the UK in the 1980s and early 1990s may also reflect some similar trends, although the absence of a growing low-skilled section of the employed labour force in Britain and in other Western European countries may reflect differences in job opportunities and welfare provision which inhibit the growth of this section of the paid labour force. In recent years we have seen the emergence of two different literatures: one on polarisation and economic restructuring and another on welfare state regimes. These literatures have rarely been brought together but, as I shall try to show, the relationship between the two is crucial.

Social polarisation: Sassen's thesis

In a well-known series of works Saskia Sassen has outlined a thesis for global cities whereby social polarisation in these cities is seen as a result of a form of economic restructuring which is particularly concentrated in such cities. This restructuring has several key elements. First, there is the shift from manufacturing to services, particularly advanced business services. This shift is seen to result in a polarised occupational and income structure which is characterised by growth both at the top and at the bottom end by virtue of the more polarised nature of the service sector, and the contraction of the manufacturing sector which contained more skilled manual, middle-income jobs.

Second, Sassen points to the growth of low-grade service jobs which are seen to be dependent on the growth at the top end of the occupational and income structure. These jobs are concentrated in the personal service sector and provide services for the wealthy. Third, Sassen points to the growth of informalisation within what remains of the 'downgraded' manufacturing sector. These low-skilled, low-paid jobs are strongly concentrated in the immigrant labour force, which is seen to be attracted to global cities by virtue of the growing labour market opportunities. This growing occupational and income polarisation is said to be linked to a growing geographical polarisation as the top and bottom of the occupational and income structure become increasingly differentiated in space.

Sassen outlined the basics of her polarisation thesis in 1984 in a paper 'The new labor demand in global cities' (Sassen-Koob 1984). She argued that economic restructuring has had several major implications in global cities. She suggested that new forms of economic growth no longer produce the type of jobs that 'were constitutive of the massive expansion of the middle class in the post-World War II period', and she argued that there is polarisation in the occupational structure, including what she terms 'a vast expansion in the supply of low-wage jobs and a shrinking supply of middle-income jobs' (1984: 139). She quoted Stanback and Noyelle's finding that

> for the services as a whole, the important observation is that there tend to be heavy concentrations of employment in better than average and in poorer than average jobs. In contrast, in manufacturing and construction the distributions are more heavily weighted toward medium- and above-average income jobs.
>
> (1982: 133)

In her book *The Global City* (1991), she argues that the evolving structure of economic activity in global cities has 'brought about changes in the organisation of work reflected in a shift in the job supply and polarisation in income distribution and occupational distribution of workers' (1991: 9). Sassen summarised her thesis as follows:

> new conditions of growth have contributed to elements of a new class alignment in global cities. The occupational structure of major growth industries characterised by the locational concentration of major

> growth sectors in global cities in combination with the polarised occupational structure of these sectors has created and contributed to growth of a high-income stratum and a low-income stratum of workers.
> (1991: 13)

She suggests that the process of economic restructuring and polarisation are common to all global cities, arguing that New York, London and Tokyo 'have undergone massive and parallel changes in their economic base, spatial organisation and social structure' (1991: 4) and she suggests that 'transformations in cities ranging from Paris to Frankfurt to Hong Kong and São Paulo have responded to the same dynamic'. These are bold claims but not everybody accepts them. I want to argue that this thesis, while very stimulating, is simultaneously partial and over-generalised by virtue of its dual focus on (a) economic restructuring and (b) the economically active labour force. Polarisation is seen as a direct and unmediated consequence of economic restructuring in global cities.

I want to argue that, while there is undoubted evidence of growing income polarisation in some capitalist economies, and particularly in the major cities, the evidence on the occupational structure of the economically active labour force points towards growing professionalisation, rather than polarisation. This is not to say that major changes in employment structure and opportunities are not taking place, but that in many Western European countries, they are more likely to create a large and growing unemployed and economically inactive group excluded from the labour force rather than the growth of a large, low-skilled and low-paid labour force. While this may be true in the USA, with its large and growing immigrant labour force, willing to work for low wages (possibly forced to because of the limited nature of welfare provision), it is not necessarily true of all western capitalist countries.

What is missing from Sassen's treatment of polarisation is that the causes of polarisation may be multi- rather than monocausal, that polarisation may be growing in terms of income but not occupation, and that the extent of social and spatial polarisation in any country may be linked to the form of welfare state in different countries. What I want to argue is that Sassen's model of polarisation is US-based, primarily monocausal, and fails to appreciate that the economic pressures towards polarisation are mediated by different welfare state regimes. What may be happening in some cities in the USA is not necessarily happening in similar cities in other counties because of differences in the economic, social and institutional context.

Income polarisation: the empirical evidence in Britain

There is strong evidence that income inequality has grown considerably in Britain, the USA and several other western countries during the 1980s and early 1990s (Buck 1994, Atkinson 1993, Jenkins and Cowell 1994, Gardiner

1993, Stark 1992). Without going into detail, a variety of data sources shows that the distribution of household income became markedly more unequal. In Britain, the top decile's share of total income rose dramatically in the 1980s while the shares of the lowest 70 per cent decreased, particularly those of the lowest 30 per cent. This is also true of income from employment as well as total income.

A specific examination of income inequality in London shows that the shares of the top decile of household incomes rose from 24.8 per cent of the total in 1979/80 to 28.3 per cent in 1985/6 and to 33.5 per cent in 1989/90. The share of the next decile rose only marginally (from 16.1 per cent to 16.5 per cent) whilst that of all other deciles fell. A similar, though less marked, pattern was found in all the other regions. Looking at the changing percentage of households in each region in the national top 10 per cent and bottom 10 per cent of gross household normal weekly income in 1979/80 and 1989/90, Stark (1992) found that the percentage of households in London in the top 10 per cent of national incomes rose from 14.3 per cent in 1979 to 20.1 per cent in 1989 while the proportion in the bottom 10 per cent also rose from 8.4 per cent to 10.3 per cent. This clearly indicates that households in London became more polarised in income terms relative to the national average during the 1980s. The degree of income inequality between rich and poor rose sharply and the proportions of rich and poor also increased.

Clear evidence of growing income inequality in London is also presented by Buck (1994), who shows, using FES micro data sets, that whereas the interdecile ratio between the incomes of the lowest and the highest decile in London and the UK was very similar in 1978–80 at 3.85 and 3.75 respectively, the ratios had risen to 8.17 and 5.94 in 1989–91. In London the interdecile ratio more than doubled in a decade. But the data on the socio-economic composition of the economically active workforce in London show no evidence of occupational polarisation (Hamnett 1994b). On the contrary, they show professional and managerial groups have grown while the manual groups have decreased.

We have, then, what in terms of Sassen's thesis is rather a paradox: growing income polarisation but no polarisation of the occupational structure. Similar evidence from the Netherlands (Hamnett 1994a) suggests that Britain is not unique in this respect, and preliminary evidence for Paris (Preteceille 1995) agrees. Indeed, there is strong evidence that, despite the growth of part-time service jobs, the occupational structure of the United States is becoming increasingly managerial and professionalised (Wright and Martin 1987). This interpretation is supported by Esping-Andersen (1993), who argues that occupational upgrading is inherent in the post-industrial trajectory of the United States. Kloosterman (1994: 171) adds:

> During the 1970s and 1980s, a decoupling seems to have taken place between the occupational level and the wage level in the United States. . . . According to Esping-Andersen, a depolarisation of the occupational

> structure has been accompanied in the US by a polarisation of wage structure.

To the extent that this is true, and professionalisation is not merely a statistical artefact of the devaluation of job titles which is accompanied by a downwards shift in work skill (a question Wright considers but dismisses), then the paradox of growing income polarisation without occupational polarisation may not just apply to Britain and other similar countries, but to the USA as well. How and why has this come about and what light does it shed on Sassen's thesis of economic restructuring as the major driving force of polarisation? Buck notes that

> Global cities arguments focus on polarisation within occupational hierarchies, and would argue that these are what is specific to large cities. On the other hand one might expect changing patterns of employment and inactivity, as well as the relative incomes of these groups to the employed population, to depend largely on state policy and national economic performance.
>
> (1994: 6)

Economic restructuring, welfare regimes and social polarisation

Sassen does touch on the issue of welfare state policy. She states in the introduction to *The Global City* that one of the three key questions it addresses is 'what happens to the relationship between state and city under conditions of a strong articulation between a city and the world economy' (1991: 14). Her answer to this question is clear. She argues that 'the nation state is becoming a less central actor in the world' (1991: 167), and that the welfare state is less important under the 'new economic regime' (1991: 338). She argues that

> There has been a generalized dismantling of a system that provided a measure of job security, health benefits, and other components of a social wage to a critical mass of workers. . . . This development raises a number of questions about the intersection of economics and politics and the state and about the 'natural' tendencies of capitalist economies. . . . is what we are seeing today – increased economic and social polarization – the 'natural' outcome of the operation of the economic system when political claims carry little weight?
>
> (1991: 333–4)

Sassen is correct that there has been a tendency towards the dismantling of the social welfare system in a number of capitalist countries but I would question that there has been 'a generalized dismantling' or a growth of income polarisation in all countries. The significance of the welfare state has varied considerably from country to country, as does the extent and the pace of the retreat from the welfare state. In order to appreciate the causes of variations in social polarisation it is necessary to focus on differences in welfare state regimes, including the availability and level of social benefits, the extent of collective consumption such as education, health and child care, state labour market intervention and the like.

Economic restructuring is extremely important as a major force shaping the nature of western capitalist countries, but economic restructuring does not occur in a social and political vacuum. On the contrary, it everywhere and always takes place within the context of nation-states with different regulatory regimes, legal structures and welfare policies and with different national and local cultures. The outcomes of global economic restructuring are essentially variable, depending on the ways in which restructuring processes are mediated within different states. Different structures of welfare and labour market practices will cushion the impacts in different ways. As Kloosterman points out:

> Descending from global logic to national and local variation . . . does not necessarily imply that the outcomes of economic restructuring will be identical in every advanced city. Differences in the economic base, the national institutional framework and urban policy have contributed to a variety of urban experiences in the 1980s.
>
> (1996: 1)

It has long been recognised in the social policy sphere that variations in state welfare structures and policies are important in shaping the patterns of social outcomes. Hilary Silver (1993) suggests that international economic restructuring has led to similar changes in employment and income structure in Britain, France and the USA. There is a shift from manufacturing to services, the absolute share of credential workers increased, more jobs became more insecure, organised labour declined in membership, female labour force participation rates rose, regional employment opportunities diverged and unemployment, income inequality and poverty all increased. But she argues that, 'While all these trends move in the same direction, they also proceed at different rates in different countries, so that there has been no national convergence in social variations in socio-economic trends.' She quotes Howell as saying: 'it is almost a truism of comparative political economy that economic imperatives create basic constraints on states but do not determine the way in which states must deal with those constraints' (1993: 339).

Silver goes on to argue that states indirectly shape the social structures through trade and industrial policies such as labour law and industrial relations policies that influence national labour markets. She instances growth of a flexible workforce in the USA and UK compared to France, where the Labour Code retarded its development. She also argues that states directly influence social structures through welfare programmes and tax structures which redistribute wealth. Silver concludes by pointing out that

> In recent years, the globalization of markets and rising unemployment have led many to believe in the economic impotence of nation-states. . . . [But] nation-states remain a vital mode of economic organization within the emerging international division of labour. . . . By modifying common global forces in product and labour markets and through redistributive welfare policies, nation-states continue to vary in their social structures. While undergoing similar social trends, national convergence is muted.
>
> (Silver 1993: 344)

A similar argument is also put forward by Lash and Urry (1993) regarding social polarisation. They argue that the process of 'reflexive accumulation' in capitalist society which gives rise to a large service class and a smaller working class also gives rise to a new lower class. But, importantly, they also argue that

> these kinds of structural change do not necessarily lead to the development of an underclass or even the creation on a large scale of a new lower class. Instead, the extent of underclass and new lower class formation depends on a deficit of institutional regulation economy and society. By contrast in societies such as Sweden and Germany, different institutional set-ups lead to a different outcome, that is to a preservation of quite a large working class and a much more limited development of a new lower class.
>
> (1993: 146)

Esping-Andersen's (1990) work on welfare state regimes is well known. He argued, on the basis of labour and social security policies, that it was possible to group countries into three major types depending on the character of their welfare regimes. Esping-Andersen's more recent work on comparative changes in class structure and mobility in different welfare regimes takes this argument further, to argue that 'contemporary social stratification is heavily shaped by institutions, the welfare state in particular' (1993: 1).

The argument advanced by Esping-Andersen and his colleagues is that the welfare state revolutionised the structure of the labour market and labour market behaviour. They argue that the social wage violates the assumption that classes and life chances can be identified through common labour market conditions in that welfare states introduce the possibility of a 'welfare state client class' (1993: 19). They also argue that the sometimes massive 'expansion of welfare state employment implies not only new occupational groups, but also the emergence of a huge production and reward system isolated from the operation of market forces' (1993: 19).

They argue that this expansion manifests itself not just by education and training programmes, but by labour market measures and direct welfare state employment growth. They note that in Scandinavia, welfare state employment growth accounts for 'almost the entire net employment increase over the past decades' (1993: 19). They also note that welfare state institutions dictate the choice of non-entry via the provision of a social wage and tax and service treatment of households and their impact on child care and working women. Finally, they suggest that the welfare state also furnishes the basic means for labour market exit through the introduction of early retirement, which provides many redundant workers with the option of a social wage rather than having to move into bottom-end jobs. In sum:

> the structure of the welfare state is a key feature in the contemporary process of social stratification: it creates and abolishes 'empty slots', it helps decide who fills them and how they are to be rewarded, it defines what is undertaken within them, and, finally, it shapes the pattern of mobility between them.
>
> (1993: 20)

The principal implication of this analysis is that there are marked differences between the occupational and income structures of different western capitalist countries. They argue that, while the Scandinavian countries 'exemplify an extreme case of a gendered, welfare state service-led trajectory', Canada and the United States are characterised by their 'large low-end consumer service labor market' (1993: 4). They suggest that

> In North America, unskilled service jobs tend to be very poorly paid, predominantly filled by youth and immigrants, and function very much as first-entry, or stop-gap jobs. In both [Canada and the USA] a distinct low-end mobility circuit emerges which is unparalleled elsewhere: unskilled sales, clerical and service jobs appear to constitute a common job reservoir for people with low education.
>
> (1993: 5)

They add that the welfare state, the industrial relations system and education are key institutional filters for employment structuration, noting that 'If the welfare state is service intensive, it will bias service employment trends; if it provides a high social wage guarantee, the scope for a low-wage-based service economy is sharply reduced' (1993: 33).

They point out that, where trade unions are centralised, strong and bargain nationally, as in Germany and Scandinavia, the employment outcome will be very different from North America where unions tend to be weak, fragmented and localised:

> the growth of low-wage service jobs is unlikely where trade unionism is comprehensive, centralized and committed to wage-equalization. In contrast, American-style industrial relations reinforce labor market dualisms and segmentation; and since trade unionism penetrates only marginally into the services, the growth of a cheap labor force is made possible. The result is likely to be more low-productivity jobs at low pay.
> (1993: 3–4)

Esping-Andersen and his colleagues undertook a comparison of trends in class structuration in six countries in the 1960s and the 1980s: Canada, Germany, Norway, Sweden, the UK and the USA. They note that, at one extreme, the USA and Canada (and, to a lesser degree, the UK) have relatively residual welfare states, characterised by only a low social wage guarantee and by a passive approach to full employment policy.

> At the other extreme, both Norway and Sweden feature welfare state and industrial relations institutions that are explicitly designed to influence the employment structure. The social wage guarantee is extraordinarily high, thus reducing the individual's compulsion to accept unattractive jobs; both welfare states feature a strong commitment to collective services, thus directing employment towards welfare state service jobs. In both cases we find comprehensive and centralized trade union systems which, for decades, have pursued solidaristic wage bargaining policies aimed at minimizing earnings differentials.
> (1993: 35)

They argue that Germany represents a third regime with a welfare state which is both generous and comprehensive but where social rights are strictly tied to employment record and there is a strong commitment to preserve the traditional caring functions of the family. Thus, there is an implicit discouragement of female labour market participation.

Esping-Andersen argues that one of the characteristics of welfare states is that they tend to produce a relatively large 'outsider surplus population', consisting of people unable to enter into employment, of early retirees, long-term unemployed and others subsisting on the social wage.

In countries such as the United States, on the other hand, where the welfare state is weaker, there is a large, low-wage service proletariat. Thus, post-industrial societies can experience two alternative kinds of polarisation. In the strong welfare states the polarisation is between 'a small, but highly upgraded insider structure and a large outsider surplus population. In the other case, a large service class proletariat will constitute the pivotal source of polarization' (1993: 28).

They were only able to study the occupational earnings structure for Germany, Sweden and the United States, but they state that

> the pervasive low-wage effect on job trends within the American consumer service sector stands out clearly. The American earnings distribution is almost the exact opposite of the German. Instead of an extraordinarily privileged top, the United States is characterized by its extremely underprivileged bottom. The unskilled service workers are . . . a very badly paid workforce.
>
> (1993: 50)

They conclude by arguing that

> The fordist hierarchy has everywhere experienced a marked decline of the traditional manual working class; to a degree this has been offset by a modest rise in clerical and sales occupations. Fordism is, so to speak, becoming post-industrialized. . . . Despite the divergent shape of the post-industrial hierarchy, there is very little evidence to suggest strong polarization. Everywhere, the trend favours higher grade occupations such that the shape of the post-industrial occupational hierarchy is biased towards the top and the middle rather than the bottom.
>
> (1993: 53)

Immigration, low-wage jobs and polarisation in the USA

It is clear from this analysis that the United States is distinctive in the importance of its low-skilled and low-paid consumer service jobs and Esping-Andersen notes that minorities occupy a disproportionate share of these jobs. The role of ethnic minorities in polarisation receives much attention from Sassen (1991: 299–317). She states that 'It is impossible to disregard the facts of race and nationality in an examination of social and economic processes in New York. To a lesser extent this is also the case with London' (1991: 299). Sassen adds that 'the large influx of immigrants into the United States from low-wage countries over the last fifteen years cannot be understood separately from this restructuring'.

Sassen is quite correct in this, that the problem concerns the nature of the causal process. As I have argued elsewhere (Hamnett 1994a), Sassen argues that the expansion of low-wage jobs has led to the growth of immigration. 'The expansion in the supply of low-wage jobs generated by the major growth sectors is one of the key factors in the continuation of the ever-higher levels of the current immigration' (1991: 316).

In other words, economic restructuring has led to changes in labour demand and migrant flows. She is opposed to the reverse explanation, namely that the existence of large-scale immigration from low-wage countries has enabled the growth of low-wage service jobs. She argues that advanced capitalism 'may promote conditions for informalization. . . . The presence of large immigrant communities . . . can be seen as mediating in the process of informalization rather than directly generating it' (1991: 282) and 'it is the economy rather than immigrants which is producing low-wage jobs'.

I have considerable sympathy with this argument, partly for political reasons, in that it is easy to slip into a right-wing, anti-immigrant stance, blaming immigrants for undercutting wages. I find it very difficult, however, to overlook the fact that the United States, particularly the major cities, has a level of immigration unparalleled in most other western countries, and that it is also almost the only western country to have seen the rapid growth of large numbers of low-wage service jobs.

Many of these jobs are common to all western cities, particularly global cities, but New York and Los Angeles are perhaps unique in the ease with which it is possible to hire cheap immigrant labour to undertake low-paid consumer service jobs. In part, these workers have no other option. The welfare benefits available in the Netherlands, Scandinavia, Germany and, to a lesser extent, in Britain are simply not available at the same level in the USA. To this extent, occupational polarisation in the USA may be the product of both the nature of the welfare state and a high level of immigration from low-wage countries.

Conclusions

The core of my argument is that while Sassen may be right that, in New York and Los Angeles, there is occupational and income polarisation within the paid labour force, the evidence from other western cities does not support this. In most large non-American cities, income polarisation is combined with professionalisation of the paid labour force. This is not to suggest, of course, that large and growing numbers of people are not experiencing low incomes. Rather it is to argue that, in countries with stronger welfare states, a larger proportion of the population may be able to live outside the paid labour force on state benefits. I accept, of course, that in a growing number of western countries the welfare state is under grave threat. We may be facing the sort of future currently found in New York and Los Angeles. But, this is not simply a result of the unmediated operation of market forces and privatisation. It is also an outcome of the historical legacy and also of political struggle to preserve welfare states against neo-liberal economic pressures.

3

ASSIMILATION AND EXCLUSION IN US CITIES

The treatment of African-Americans and immigrants

Susan S. Fainstein

The United States has a long and contradictory history in its absorption of groups defined by race and ethnicity. Along with Canada and Australia, it is among those few settler states where colonists rapidly became the numerical majority. But it also differs from other economically developed nations in having been, for a substantial period, a slave-holding society. As a consequence, social dominance and foreign birth do not necessarily clash, but social dominance and dark skin colour do. In order to understand the American system of inclusion and exclusion, we must distinguish between the experiences of immigrant groups and Blacks and also understand how their differing histories influenced each other.

The creation of American culture

American politics and society have been defined by the early triumph of settlers from Europe over indigenous inhabitants, the eventual consignment of Native Americans to reservations (the original 'other'), the subsequent absorption of waves of immigrants from a multiplicity of homelands, and the exclusion of Black Americans from the full benefits of citizenship. Hence, the story of spatial segregation and social exclusion in American cities results from the ideological legacy and unstable interaction of two phenomena: (1) voluntary immigration, which created successive tides of newcomers, who were self-selected and who mainly could eventually fade into the general population; and (2) involuntary servitude of a physically identifiable group, establishing a caste that even after emancipation could not escape stigmatisation. As a result, the social construction of otherness takes on varying forms in American cities and has undergone a number of temporal changes.

The history of group inclusion and peripheralisation, rather than class relations, has, in fact, been the defining matrix of US political traditions and social relations. Americans do not possess the kind of strong national culture, so evident in European nations and Japan, that makes foreigners stand out among their populaces. The nineteenth century in Europe marked the consolidation of national cultures, the ebbing of defiance of the national state by indigenous cultural/territorial groupings, and the rise of class-based movements and parties. Suppression of internal movements for political and cultural independence required the imposition of a hegemonic national identity. In contrast, American national consolidation was established in the face of challenges from foreign-born rather than internal forces.[1] By the latter part of the nineteenth century the central tension in politics, especially urban politics, manifested itself in the contest between the political machine, which mobilised voters through ethnic networks, and the White Anglo-Saxon-dominated progressive and populist 'reform' movements. The outcome of the struggle produced a looser conception of nationhood than exists in Europe. As de Tocqueville pointed out, American political culture encompassed a weak state and a highly pluralist civil society. This civic context combined with a lesser emphasis, as compared to Europe, on place of birth and cultural practice as the emblems of national identification. In addition, an individualist ethic minimised the importance of each citizen feeling him or herself to be part of a national community with mutual obligations.

Altogether these characteristics, as well as a long tradition of recruiting foreign labour for low-wage or otherwise undesirable jobs, have contributed to a highly permissive legal framework for the transformation of foreigners into citizens. The Chinese Exclusion Acts and a series of restrictive immigration acts beginning in 1924 did for a period limit access to citizenship by national origin.[2] For most of American history, however, acquiring citizenship has been relatively easy, based primarily on desire rather than ancestry, economic status, or knowledge. At the present time the United States does not issue national identity cards, and the occasional proposals that it do so are vehemently resisted. Moreover, except in California and along the Mexican border, there has been only a limited coincidence between ethnicity and territory, sharply restricting the potential for autonomy movements.

In a country that congratulates itself on being 'a nation of immigrants', ethnic differentiation itself becomes a cultural tradition and a source of legitimation. The words inscribed on the Statue of Liberty sum up the positive ideological spin that surrounds immigration: 'Give me your tired, your poor, your huddled masses yearning to breathe free.' They indicate the acceptability of two motives for coming to the USA – the desire for economic improvement and for political freedom. Various conservative commentators have used the material success achieved by some ethnic groups – especially Jews, (Asian) Indians, Chinese and Koreans – to substantiate their pro-market, anti-welfare philosophies and, by implication, to blame Black economic failures on lack of initiative. Candidates for mayor of New York City typically visit the countries of origin of its largest population groups (traditionally 'the three I's' – Italy, Ireland and Israel). The current Republican mayor, Rudolf Giuliani, lauds the benefits of immigration as heartily as his Democratic predecessors. Recently his deputy mayor commented:

> Mayor Giuliani believes that it is very wrong to restrict immigration. Immigrants are what make this city so vital. We say that regardless of your legal status, once you get here you are part of us. We remain today the great recipient of this world's people.[3]

Nevertheless, Americans are not consistently welcoming of immigrants. Sporadic, sometimes deadly assaults on foreigners receive little publicity and are not associated with political movements as in Europe, but occur nonetheless. The multiculturalism so frequently celebrated in a host of civic events and 'UN tickets' of political candidates at times has been deemed a dire threat to common purposes and communal identity. At the present moment some dissonant voices are making themselves heard in a period when the United States is witnessing the highest rate of immigration since the turn of the century.[4] In particular, in California, the state which has far exceeded all others as a recipient of immigrants, immigration became the central issue of a recent political campaign. A citizens' initiative had placed on the ballot a proposition that would prevent undocumented aliens from entering their children in school and would deny them access to various social services.[5] Although the Democratic candidates for state-wide office opposed the measure, their Republican opponents were vociferous in its support. An indication, however, of the recency of this position is the fact that Pete Wilson, the successful Republican gubernatorial candidate, had earlier been a sponsor of an act, passed while he was a US Senator, that exempted Mexican farm labourers from immigration restrictions.

African-Americans differ from voluntary immigrants and their descendants both in their own attitudes towards the majority society and in the way that majority views them. Although immigrants often maintain ties with their homelands, establish a variety of mutual support schemes, and participate in ethnic associations, they have not developed a separatist ideology. In fact, ethnically identified Americans are often the stalwarts of patriotic organisations and the most effusive in praise of 'the American way of life'. In contrast, most African Americans do not regard the United States as a land of opportunity. The experience of slavery, the snatching away of the gains of Reconstruction through the imposition of Jim Crow, and the denial of civil rights, as well as continued housing and employment discrimination and economic deprivation, have caused them to regard the United States as hostile territory. Their pessimism has in widely publicised instances stimulated the development of a separatist, Black nationalist ideology that exacerbates isolation and, less blatantly, of an alienation that manifests itself in what potential employers frequently regard as 'an attitude' – i.e. a lack of responsiveness and minimally disguised anger. In turn, the mainstream perception of Blacks as practising 'reverse racism', along with resentment at governmental affirmative action programmes, further hardens negative White opinion towards them.

Despite – and because of – the peripheralisation of Native Americans and Blacks, and even in the face of periodic successes by (White) nativist movements in arousing anti-immigrant sentiment, America has been remarkably successful in assimilating numerous groups that in other national contexts would have remained in social isolation. At the same time the failure of class-based mobilisations has reinforced the economic disadvantage of people of colour and of those ethnic groups that have been left behind. And because those economically worst off can often be identified in racial and ethnic terms, Americans up and down the social hierarchy largely locate the causes of poverty in group attributes rather than class exploitation. A closer examination of the different situations of immigrants and their descendants and of African-Americans reveals the forces that create the dynamics of exclusion, inclusion and socio-economic status in US cities. Finally, a comparison of the US and European definition and treatment of 'otherness' offers insights into the relationship between social difference and the impacts of the welfare state on inequality.

Immigrants in US cities

American immigration legislation reduces the likelihood that newcomers to the United States will arrive as outsiders. During the period of greatest restriction on immigration, individuals were admitted according to national origins proportionate to their representation in the 1890 Census, thus ensuring that those who came would resemble those already there. Even after 1965, as a series of immigration laws liberalised the national origins system to give equal access to people of all nations and greatly raised the aggregate permitted to enter the country in any year, family reunification remained the first priority for determining eligibility. Consequently, most immigrants form part of already existing communities and have blood ties to people previously living in the United States. Except as a part of certain agricultural labour importation schemes, a system of hostels or special housing estates for foreign workers and a view of immigrants as only temporary sojourners are virtually absent.

According to Sassen (1988: ch. 4), the stimulus for migration flows results from the disruptive effects of direct foreign investment on the social structures of less developed countries. She further contends that the choice of destination by the migrants results from the foreign presence in their place of origin. Thus, extensive direct foreign investments by American companies have increased the flow of immigrants to the United States through giving the potential immigrant pool some familiarity with American culture. In the case of professionals, an increasingly significant category of migrants, many of whom come as students and may not initially have intended to stay, this preliminary acculturation is enhanced. Once arrived in the United States, as compared to Europe, immigrants from the less developed countries do not confront tensions resulting from the inheritance of a colonial past. US world economic power and the anti-Americanism it has provoked thus do not translate into the personal tensions that derived from the colonial relationship (see Mannoni 1956).

Immigrants to the USA predominantly settle in cities. In the years 1961–90 more than 15 million legal immigrants entered the country, nearly half of them in the 1980s. During the same period 4.8 million foreigners became naturalised US citizens.[6] They particularly gravitated to a relatively small number of metropolitan areas, with more than half of the 1990 total settling in only four metropolitan areas (see Table 3.1). For obvious reasons they selected those places that had enjoyed economic expansion during the 1970s and 1980s; the cities of California, which both shares a common border with Mexico and enjoyed unprecedented prosperity during that period, attracted close to 40 per cent of total immigration in 1990.

The southern California circumstances resemble the situation in certain European countries. In California, as in southern France and in Italy, the closeness of the immigrants' home country reinforces the strength of native culture. The recency and relative homogeneity of the immigrant group also make the California situation more like Europe (see Tables 3.2 and 3.3). Furthermore, as in Europe, much of the receiving population feels that the newcomers represent a sharply contrasting culture. This perception reflects the fact that California, before the recent influx of immigrants, differed from the east coast states in being heavily White Protestant and not having a tradition of immigrant assimilation.

Table 3.1 Immigrants admitted to USA, by metropolitan area of intended residence, 1990

Metropolitan area of intended residence	*Number*	*% of total*
LA/San Diego, CA region[a]	512,964	33.4
New York, NY	164,330	10.7
San Francisco, CA region[b]	76,288	5.0
Chicago, IL	73,107	4.8
Houston, TX	58,208	3.3
Miami, Hialeah, FL	37,677	2.4
Washington, DC/MD/VA	32,705	2.1
Dallas, TX	28,533	2.0
Total, all metropolitan areas	1,536,483	

Source: US Bureau of the Census (1992) *Statistical Abstract of the United States: 1992*, Washington, DC: US Government Printing Office, Table 11

Notes

a Includes Los Angeles/Long Beach, Anaheim/Santa Ana and Riverside/San Bernadino metropolitan areas.

b Includes San Francisco, San Jose and Oakland metropolitan areas.

Table 3.2 Racial breakdown for the city of Los Angeles

	1960	*1970*	*1980*	*1990*
Total pop. (mil.)	2,479	2,816	2,967	3,485
% Black	14	18	17	14
% Hispanic	na	na	28	40
% Asian	na	na	na	10

Sources: US Bureau of the Census (1976) *Statistical Abstract of the United States*, Table 23; 1985: Table 26; 1994: Table 46

Note
The figures for Hispanic and Asian were not given for 1960 and 1970 and for Asian also not given for 1980 and thus were collapsed into both Black and White categories in those years.

Table 3.3 Country of origin of immigrants to Los Angeles and New York City areas, 1990

Area/top countries of origin	*No. of immigrants*
Los Angeles area	
Mexico	866,088
El Salvador	169,077
Philippines	114,123
Vietnam	91,442
Korea	91,385
New York City area	
Dominican Republic	145,153
China	74,959
Jamaica	74,168
Colombia	66,685
Korea	58,456

Source: New York Times, 30 October 1994: section 4

Nowhere in Europe, however, does the magnitude of immigration begin to approach southern California levels.

In New York, political structures and school systems are reverting to the methods of a now mythologised past, catering to ethnically defined voting blocs and dealing with as many as fifty different languages in a single school. (Tables 3.3 and 3.4 show the composition of the New York population.) Bilingual education remains a hotly debated topic, but many of the immigrants' children are regarded as more highly motivated than those of African-Americans and thus are welcomed by many educators. In New York the recent City Charter frankly recognises ethnic difference and calls for the delineation of voting districts in a fashion that will encourage representation of ethnic and racial groups in the city council in proportion to their share of the electorate. The varied background of the immigrant stream has meant that no particular

Table 3.4 New York City population (000s)

	1940	*1950*	*1960*	*1970*	*1980*	*1990*
New York City	7,455	7,892	7,782	7,895	7,072	7,323
% White	94	90	85	na	61	52
Manhattan	1,890	1,960	1,698	1,539	1,428	1,488
% White	83	79	74	na	59	58
Brooklyn	2,698	2,738	2,627	2,602	2,231	2,301
% White	95	92	85	na	56	47
Bronx	1,395	1,451	1,425	1,472	1,169	1,205
% White	98	93	88	na	47	36
Queens	1,298	1,551	1,810	1,986	1,891	1,952
% White	97	96	91	na	71	58
Staten Island	174	192	222	295	352	379
% White	98	96	95	na	89	85

Sources: L.C. Rosenwaike (1972) *Population History of New York City*, Syracuse: Syracuse University Press, 121, 133, 136, 141, 197; US Bureau of the Census (1986) *State and Metropolitan Area Data Book*, 202, Table A; US Bureau of the Census (1990) *Census of Population and Housing, Population and Housing Characteristics, New York, NY PMSA*, 734–5

Note

Individuals of Hispanic origin are included in both White and non-White categories depending on how they identified themselves. The city contained 21 per cent of Hispanic origin in 1980 and 25 per cent in 1990. Percentage White is not supplied for 1970 because the treatment of those of Hispanic origin in that year differed from the rest of the series. Groups not included in the 'White population' are Blacks, Native Americans, Asian/Pacific Islanders and 'others'.

group overwhelms the others, thus lessening fear of domination and dampening any movement towards autonomy.

Waldinger delineates the complex process through which immigrants positively affect urban economies, claiming 'had there been no immigration, New York would have suffered an even more severe [economic] decline'. In a paper written with Cross (1992: 162–4), he argues that out-migration of the White population from New York exceeded the outflow of jobs, thereby leaving vacancies even in the face of employment decline. He traces a sequence of events whereby the initial influx of immigrants replaces outmigrating native workers; this new, lower-wage labour force then permits businesses to compete more successfully in international markets and increase the demand for locally produced goods and services. These outcomes, in turn, lead to the establishment of new business niches and support systems and increase the absorptive capacity of the local economy for additional immigrants. As well as reinvigorating the formal economy, immigrants form the base of a developing informal sector that contributes to the further extension of global city functions, albeit often at the price of extreme self-exploitation (see Sassen 1994b).

Analysis of census data reveals impressive levels of inter-generational upward mobility among some immigrant groups and diversity in the earnings of different immigrant groups (N. Fainstein 1993).[7] New York City Asians born in the

United States are disproportionately in the top 40 per cent of earners, as compared to overseas-born Asians, of whom about 55 per cent are in the lowest two quintiles. Overseas-born Whites show earnings levels exactly proportionate to their share of the population. About 65 per cent of overseas-born Hispanics are in the bottom 40 per cent of earners, while 49 per cent of those born on the mainland are similarly impoverished. Black immigrants in New York are similar to Asians, with about 54 per cent located in the bottom two quintiles of earners, making them better off than their Hispanic counterparts.[8] In Los Angeles, however, Hispanic immigrants who identify themselves as Black fare considerably worse than those who consider themselves White (Logan *et al.* 1994: 129). On the whole, the new immigrants contribute to the strengthening of urban economies and the revitalisation of decayed neighbourhoods. Without immigration the major recipient cities – the 'gateway cities' in Muller's (1993) phrase – would have suffered substantial population decline and continued neighbourhood deterioration. The impact of these new residents on depopulated and decayed parts of the city is immediately obvious:

> Outside the yuppie strongholds of Manhattan and other favoured areas in Brooklyn and Queens, immigrants have been the leading factor in neighbourhood revitalisation. Owing to their high employment rates and multiple wage earners, the new foreigners have injected large doses of new purchasing power into the rehabilitation of an ageing housing stock and the resurrection of inert retail streets. Their presence has been visible not only in the demand for housing but in the supply as well, displaying a willingness to undertake the disagreeable, if potentially profitable, tasks entailed in fixing up and managing decaying buildings. In great numbers, too, they have accepted the arduous hours (and the dangers) of operating small retail stores. Here and there, they have also restored the semblance of a long-absent night life, for example, Brooklyn's Brighton Beach and the Ironbound section in Newark.
>
> (Winnick 1990: 62)

At the same time, however, as immigrants contribute to economic activity and neighbourhood revitalisation, they depress wages and diminish working conditions. Consequently, while they keep down the rate of inflation and increase urban industrial competitiveness, they also contribute to heightened inequality and the degradation of work. Moreover, as they have come to dominate particular low-skilled occupational niches, they have caused African-Americans to be excluded from them.

Immigrant enclaves differ from Black ghettos in that their impetus is mainly voluntary, although the skin colouring of the ethnic group affects its ability to disperse. At all income and educational levels immigrants are less spatially isolated than African-Americans, despite the expressed willingness of the latter group to live in neighbourhoods in which they are a minority (Jaynes and Williams 1989: ch. 3, Massey and Denton 1993: ch. 4). American ethnic enclaves diverge also from the immigrant ghettos of Europe in that their occupants are primarily in the private housing market. The likelihood of a single nationality dominating publicly owned buildings – as Surinamese do within Amsterdam's Bijlmermeer, Algerians within some of the HLMs of Paris, or Bengalis in London council estates – is virtually non-existent in American cities. Instead, the scarcity of social housing in the USA and the still substantial private rental sector have meant that immigrants with little income cluster in areas of privately owned low-rent housing. Their landlord is often a fellow countryperson or, more recently, a community development corporation that owns scattered, small-scale buildings.

Americans in the past blamed immigrants for the degradation of their cities and their politics. Today, despite the very high levels of immigration into American cities, Blacks, rather than foreigners, are the feared 'other'. Multiculturalism, although often derided as 'politically correct' and destructive of the common culture, permeates educational curricula and public rhetoric. Ethnic associations and religious congregations identified by the national origins of their members meld easily into the fabric of American life. The degree of economic success and social acceptance achieved by many immigrants puts in sharp relief the continued social exclusion felt by African-Americans.

African-Americans in US cities

African-Americans have endured well over a century of persistent discrimination since the Emancipation Declaration. Social and economic disadvantage have interacted in a pattern that Gunnar Myrdal (1944) named 'cumulative causation'. Although African-Americans have made substantial economic gains in the 1950s and 1960s, their position relative to Whites has not improved since then. Likewise residential exclusion has diminished little. Even though many Blacks have followed Whites to the suburbs, it is important to remember that their suburbs are not the same as those of White occupation (N. Fainstein forthcoming).

White rationalisation of the isolation of African-Americans has become less overt and is no longer supported by law. It continues, however, to receive 'scientific' justification. There are currently two popular interpretations, one environmental and one genetic – the underclass and IQ arguments – that are being used to explain Black deprivation. Both arguments put the economic deprivation of Blacks in racial rather than class terms.

The underclass thesis roots the unabated poverty and high rates of social pathology of urban Blacks in the economic and social isolation of the 'hyperghetto' (Wacquant 1994). This locale of extreme alienation, which has been abandoned by upwardly mobile African-Americans and detached from employment opportunities, lacks the communal bonds that previously sustained the poor and socialised the young (Wilson 1987). It no longer offers a sufficient pool of marriageable men (i.e. men able to earn a large enough income to support a family) to permit most children legitimate births, resulting in the absence of stable family settings. Essentially a reprise of the 'culture of poverty' thesis of the 1960s, the underclass hypothesis does not blame Blacks directly for their situation. Rather, it puts the onus on a system that results in the segregation of poor Blacks, who, in a new version of the cumulative causation argument, lack attributes that would make middle-class Blacks or Whites willing to live with them. But it identifies the immediate cause of their situation in their own traits and in the flight of upwardly mobile Blacks from inner cities rather than in the economic inequality generated by the labour market practices, social discrimination and political decisions of the larger society (see N. Fainstein 1993, Squires 1994).[9]

The underclass argument is correct in stressing the absence of opportunity in isolated Black ghettos. Unlike many ethnic enclaves, 'hyperghettos' contain few ladders of mobility resulting from internally generated businesses. The issue, however, is one of emphasis. The underclass hypothesis implies that once individual economic success allows Blacks to move into less racially isolated communities, they can compete on equal terms with Whites. Blacks in the United States, however, remain more segregated than any other identifiable racial or ethnic group at all income levels and continue to experience housing and employment discrimination even when they have escaped from poverty.[10] Only a small percentage of African-Americans belong to the underclass if it is defined by complete detachment from the labour force, yet almost all Blacks, even those with relatively high incomes, are disadvantaged when compared to Whites. Thus, as Table 3.5 shows, the ratio of Black to White earnings for full-time workers declined slightly between 1973 and 1990 and never exceeded

Table 3.5 Earnings by race, all employees and state and local government employees, 1973–90

	1973	*1980*	*1990*
All full-time employees			
Median weekly earnings ($)			
Black	129	208	329
White	162	265	427
B/W (%)	80	78	77
Full-time state and local government employees			
Median yearly salary ($000s)			
Black	7.4	11.5	22.0
White	8.8	13.8	25.2
B/W (%)	84	83	87

Source: Statistical Abstract of the United States 1980: Table 704; 1986: Table 704; 1992: Table 483

Note
Weekly earnings in 1973 are for 'Black and other races'.

80 per cent. Since, by definition, full-time workers do not belong to the underclass, the relatively low earnings of Blacks cannot be explained by the underclass hypothesis. In other words, one simply cannot understand the wide overall differences between Blacks and Whites by focusing on the situation of those at the bottom.

The IQ argument has again come to the fore with the publication of Herrnstein and Murray's book *The Bell Curve* (1994). Although heavily attacked by every respectable reviewer, this book has received enormous publicity. It argues that in the present era good jobs require high intelligence, that intelligence is genetically transmitted, and that African-Americans lack the necessary genetic traits to produce intelligent offspring. It is unlikely that the book, which is heavily larded with statistical evidence, will persuade anyone not already convinced of its argument. Its mode of framing the issue, however, as with the underclass debate, focuses public attention on the characteristics of African-Americans rather than the relationship between them and White society. And, of course, the invidiousness of its thesis and its support for saying what had been for a while unsayable will further delegitimise African-American demands for economic equality.

Circumstances have changed, however, since Myrdal wrote *An American Dilemma*, and Blacks no longer confront their situation with the passivity that Myrdal identified. In particular, as a consequence of protest movements, of the civil rights laws adopted during the 1960s, and of Black numerical strength in a number of cities, African-Americans have moved broadly into political office, particularly at the municipal level. Most large American cities have had Black mayors, while the number of Black elected municipal, law-enforcement and education officials jumped more than fivefold between 1970 and 1991, from 1,300 to 6,969; this increase occurred primarily in urban areas (US Bureau of the Census 1992: Table 431).

The capture of public office has turned out to have mixed results for African-Americans (S. Fainstein and N. Fainstein forthcoming). In terms of public policy, Black political power has succeeded in halting the massive land clearance and displacement efforts that had characterised the post-war urban renewal programmes. Black elected officials, however, have largely failed to attract the allegiance of White working-class residents and perforce have had to rely on business interests for continued electoral success and governmental effectiveness (Reed 1988). Consequently their economic development efforts have offered few specific benefits to low-income minorities and have largely followed the pattern set by White, development-oriented urban regimes (S. Fainstein and N. Fainstein 1989).

Black electoral triumphs had their greatest impact on the composition of public bureaucracies (see Eisinger 1980, Browning, Marshall and Tabb 1990, Bennett 1993). Blacks are both strongly represented in government jobs and disproportionately dependent on public-sector employment. Table 3.6 shows Black representation in government employment over two decades and compares

Table 3.6 Black representation in government jobs, 1972–90

Full-time government employment (000s)	*1972/3*		*1982/3*		*1990*	
Federal civilian except postal	1,887		2,009		2,150	
Blacks (% Black)	267	(14)	311	(15)	357	(17)
Postal service	655		666		809	
Blacks (% Black)	123	(19)	127	(19)	178	(22)
State & local gov't except education	3,809		4,492		5,374	
Blacks (% Black)	523	(14)	768	(17)	994	(18)
Public schools	3,452		3,082		3,181	
Blacks (% Black)	433	(13)	432	(14)	463	(15)
Total, all government	9,803		10,249		11,514	
Blacks (% Black)	1,346	(14)	1,638	(16)	1,992	(17)
FTYR civilian employment	54,859		63,807		80,853	
Blacks (% Black)	5,091	(09)	5,838	(09)	8,281	(10)

Sources: Statistical Abstract of the United States 1974: Table 390; 1978: Table 246; 1980: Table 477; 1991: Tables 499, 652, 915; 1992: Tables 236, 483, 522, 629, 880. *Current Population Reports*, Series P-60, No. 142, Table 37; No. 174, Table 26

Note
Employment data by race were unavailable for the Postal Service in 1982 and 1990, so an estimating method was utilised. We calculated the proportion of postal mail carriers and postal clerks that was Black from the detailed breakdown by occupations for all workers. Since these occupations account for about 80 per cent of all Postal Service jobs, we multiplied the racial composition percentage by the total number of Postal Service employees in each period to estimate the number of Black employees. The percentage Black of FTYR (full-time, year-round) workers in all sectors provides a comparison for the percentage of workers in the government sector alone.

it to overall civilian employment. It indicates that Black representation increased in every category of national and sub-national government, even during periods when the national administration was unsympathetic to affirmative action programmes. Moreover, government jobs have been disproportionately significant in allowing African-Americans to become part of the middle class. Table 3.7 compares the number of job-holders in the public sector who occupy well-paid and high-status positions with the number of individuals in each racial group who have relatively high incomes – i.e. jobs paying $25,000 or more in 1980 dollars. The table clearly demonstrates that the Black middle class depends on the public sector for its sustenance to a much greater degree than does the White. In 1990 the dependency ratio for Blacks was 1.75 times the figure for Whites.[11]

Politics and public administration thus constitute the areas in which African-Americans are most strongly integrated into American life. Ironically, however, this success results from their territorial segregation. The US system of strong, independently elected mayors, district-based city councils, and single-member Congressional districts allows African-Americans to mobilise as an areally based voting bloc for Black candidates. Dispersal would do away with their majorities.

Table 3.7 Dependence of employees with middle-class earnings on government jobs, 1980–90

	1980		*1990*	
	Black	*White*	*Black*	*White*
Persons with middle-class government jobs (000s)				
Federal GS-9 and above	58	687	93	838
Professional/management/technical occupations, state and local government	128	1,175	261	1,577
Public school teachers and administrators	206	1,759	209	1,576
Total	392	3,621	563	3,986
Persons with middle-class incomes (millions)	4.7	75.7	7.8	97.2
Index of middle-class dependence on government jobs	0.08	0.05	0.07	0.04

Sources: Statistical Abstract of the United States 1984: Table 546; 1986: Table 229; 1992: Tables 236, 522. *Current Populations Reports*, Series P-60, No. 180 (1991) Table B14

Note
The index of middle-class dependence is defined as the ratio of persons with middle-class government jobs (as specified in the table) to persons with middle-class annual incomes. The latter includes all persons 15 years of age and older with incomes of $25,000 or higher (in 1991 CPI-U-X1 adjusted dollars). In both years the White-male median income was about $22,000. Approximately 44 per cent of White males had annual income of $25,000 or higher in both years. The GS-9 pay grade began at $24,705 in 1990. The table underestimates the number of employees with middle-class government jobs to the extent that it includes only unambiguous categories. For example, it omits the GS-5–8 pay grade, which overlaps slightly with GS-9 and above. It also omits persons with government protective-services occupations, even though the median income of such is just above $25,000.

In the United States, unlike Europe, voters select candidates in primaries and vote for only one candidate for legislative elections. Consequently, if Blacks did not dominate the electorate of particular districts, they would be unlikely to achieve any representation at all.

'The other' in US and European cities

In both Europe and the United States immigration has accelerated in the last decade and a half, although Europe has not yet reached American levels. As in the USA, immigration into Europe is uneven, with disparate countries and cities receiving varying numbers of immigrants from different places. The reasons for the increase in both the USA and Europe are twofold: disruptions in various parts of the world have created streams of refugees; and the restructuring of the world economy has uprooted people from customary occupations and driven them to seek work outside their native lands.[12]

Despite this similarity of experience and the existence of definable outgroups within European countries that are correlated with racial and cultural differences, comparisons of the US and European situations with regard to racial and ethnic segregation are inexact. The difficulty lies in selecting the groups to be equated.

The much larger numbers of outsiders who have been absorbed into American society also make the comparison imperfect. The net increase in US population from immigration between 1970 and 1990 is estimated to exceed 10 million, accounting for about a quarter of overall population gain (Muller 1993: 114). During the same period Europe had only a small amount of population growth and much less immigration. Unquestionably African-Americans are both poorer and more segregated than any counterpart group in Europe, except perhaps Gypsies. But immigrants, especially Arabs and Asians, who comprise identifiable ethnic minorities in Europe, fade fairly easily into the texture of American life. Thus, it is when immigrants to Europe are matched with African-Americans that the USA appears more discriminatory, not when immigrant groups from Asia and the Middle East within the two areas are compared to each other.

Certainly the greater reach of the European welfare state means that in most EU countries outsider groups can obtain decent housing and for the most part do not live in neighbourhoods substantially lacking in services. Thus, in a study of housing policy in the Netherlands, Germany and the UK, Heisler (1994: 214) concludes that

> Compared to the United States, the spatial dimensions of relatively poor neighborhoods [in Germany and the Netherlands] are small (particularly in the Netherlands) and they are far less isolated than their US counterparts. This is partially due to neighborhood size, but also to their remaining heterogeneity and the continued presence of a social and economic infrastructure. While such neighborhoods may not be pleasant places to live, they are not devoid of a large variety of economic and social institutions (churches, stores, physicians, dentists, neighborhood associations, self-help groups, and social assistance offices). Finally they are connected to the city as a whole by good public transportation systems.

Heisler is less sanguine about the United Kingdom, where she cites a trend towards increasing concentration of poor people, but she also notes that some immigrant enclaves there, while poor, are also buoyant and heterogeneous.

A comparison of the housing situation of minorities in three cities in Germany, France and the United States by Sellers (1994) reaches similar findings. Minorities lived under worse conditions than majority groups in all three cases, but nevertheless the differential in the European cases was smaller than in the American one, where the minority groups identified were Black and Hispanic. Despite the widely publicised attacks on foreigners in Germany, Sellers discovers less support for integration of foreigners in France than in Germany. He attributes this difference to the relative lack of 'compensatory norms' in France – that is, to policies favourable to minorities that developed in response to the crucibles of Nazism.

Politicians within Europe most openly oppose immigration in the UK and in France. In these countries hostility to newcomers, while most virulent on the right, is also evident on the left. The Labour Party in England and the Communist Party in France both regard immigration as a serious danger to maintenance of the prevailing wage and thus combine with anti-welfare state forces in wishing to restrict access to public benefits. Leftists in the United Kingdom, who had been sympathetic to Muslim immigrants, were alienated by the Rushdie affair and became less supportive. Conflicts over the right of Muslim girls to wear headscarves in French schools have provoked similar antipathy among leftists there.

In the USA coded racism forms the basis for political appeals, but except in California the thrust is primarily directed against African-Americans not immigrants. Even the bombing of the World Trade Center by Islamic militants and a terrorist plot to blow up other New York sites did not produce any serious backlash against Muslims.

The United States differs from Europe in its historical reliance on geography as a method of maintaining class distance. The American normative system eschews rank as an indicator of worth, bans titles of address and other outward signs of deference, and establishes hierarchy sheerly on 'merit', which is almost solely interpreted as wealth. The wealthy have always tried to put as much physical distance as possible between themselves and other mortals, partly because when in contact with others they would be forced into ritualistically democratic behaviour.

Rising crime rates in US cities have increased the impetus towards separation. Social analysts have particularly noted the movement towards the creation of actual physical barriers between parts of the city and the development of office and retail centres that resemble fortifications (see Davis 1990, Judd 1994). The movement towards erecting physical divisions among parts of the city has proceeded faster in the United States than in Europe, but it is occurring there as well. Body-Gendrot (1994: 223) points to the development of marginalised youth cultures in European and American cities as an impetus to both flight by the upper classes and the development of a strategy of creating unrest to attract public attention by the media. Moreover, Marcuse (1994: 47–8) contends that the drive towards separation also emanates from poor people who find their neighbourhoods endangered by the inmigration of the well-to-do:

> Does the wall perpetuate power, or defend against it? . . . Does it strengthen hierarchical relationships among people, or does it pave the way towards greater equality?
>
> Morally these are, it seems to me, the most important distinctions. . . . In today's cities, the poorer residents of the Lower East Side of Manhattan, of Kreuzberg in Berlin, of the area around the University of southern California in Los Angeles, wish to keep the gentrifiers out as much as the residents of the suburbs and luxury housing of

Manhattan, Berlin, Los Angeles, want to keep them out, yet the two desires are not equivalent morally.

We see then certain tendencies common to contemporary urban society within the wealthy countries. They result from the globalisation of economies and the heightened mobility of capital and labour intrinsic to it and from international cultural trends including the creation of a youth culture. The tendencies reinforce internal divisions within societies in which class becomes conflated with ethnic and racial status. The characteristics associated with out-groups vary from society to society. But neither Americans nor Europeans are as willing to support people whom they identify as outside their community as those within. In the case of the USA, with its much weaker welfare state, many of these outsiders are physically within its territorial boundaries but excluded from the full measure of social citizenship and walled off from the rest of society. In the case of Europe, internally located excluded groups receive much greater economic and social protection. European culture, however, makes the social adjustment of immigrants less smooth than in the USA, and Europe itself is exclusionary, relying on restrictive immigration laws to keep out the impoverished masses of the less developed world.

Notes

1 One can interpret the Civil War as contradictory to this thesis. Secessionism, however, was not based on difference derived from the coincidence of region with a separate language, religious or cultural stock, as, for example, the Basques or the Scottish represent within Spain and the United Kingdom, respectively. Rather, the Civil War was a conflict between regions with two different class structures based on differing modes of production.

2 See Muller (1993) for a history of US immigration policy, the sources and manifestations of anti-immigrant sentiment, and the impact of immigration on cities.

3 Comments by Fran Reiter at Conference on 'Restructuring Urbanism II: The Next New York', New York, 22 October 1994. She was expressing the Giuliani administration's opposition to proposals that illegal immigrants be made ineligible for educational and social services.

4 A recent survey indicates that, while 31 per cent of Americans say that 'immigrants strengthen the USA with their talents and hard work', twice as many believe that 'immigrants take our jobs, housing and health care' (Poll conducted for the *Times Mirror* Center for the People and the Press, reported in *USA Today*, 2 November 1994).

5 The proposition received an electoral majority but its enforcement awaits the outcome of judicial decisions as to its constitutionality.

6 A large but unknown number of undocumented immigrants, estimated at about 3 million altogether in 1986, simultaneously arrived in the United States. Because of their legal status, they were less likely to have a political impact than those who could become citizens. The low proportion of resident aliens that have become citizens partly reflects the delay between acquiring resident status and qualifying for citizenship but also indicates a lack of interest on the part of immigrants in exercising political rights.

7 The analysis presented in this paragraph is drawn from an unpublished table prepared by Norman Fainstein showing earnings inequality in the New York City workforce in 1980 and 1990, based on US Bureau of the Census Public Use Microsamples, 1980, 1990. CUNY Center for Urban Studies, October 1994.
8 The figures given are for 1990. They differ little, however, from those for 1980, except that mainland-born Hispanics in the bottom two quintiles of the workforce dropped from 56.5 per cent to 49.4 per cent.
9 In fact, middle-class Blacks are more likely to associate with poor Blacks than are middle-class Whites with poor Whites (N. Fainstein forthcoming).
10 John Logan has calculated the index of dissimilarity for 1990 for the New York–New Jersey CMSA (personal communication). Only African-Americans, Black West Indians and Dominicans have indexes above 0.8 (0.81, 0.82 and 0.84, respectively), where a level of 0.99 would indicate that members of the group live in totally homogeneous census tracts. The next highest levels are 0.71 for Puerto Ricans and 0.61 for Chinese.
11 Because government employment as a basis for middle-class income declined by a small but equal amount for both races in the 1980s, the relative dependence of Blacks compared with Whites increased between 1980 and 1990. Thus, the index figure for Blacks stood at 0.08 in 1980 compared with a figure of 0.05 for Whites (boxed figures in Table 3.7). By these figures, the Black middle class was about (0.08/0.05 =) 1.6 times as dependent on the public sector as was the White. In 1990 the dependency ratio had increased for Blacks to 0.07/0.04. This analysis is drawn from S. Fainstein and N. Fainstein (forthcoming).
12 Sassen (forthcoming) goes so far as to contend that the language of immigration is inappropriate to describe the latter process and that the internationalisation of labour should be seen as the counterpart to the internationalisation of capital.

4

CHICAGO: SEGREGATION AND THE NEW URBAN POVERTY

Jerome L. Kaufman

Introduction

Officially incorporated as a government in 1837, Chicago has had a remarkable history as it raced past many other American cities in the brief span of a century to become the country's second largest city, the industrial heart and transportation hub of the nation.

Two significant decisions were made by the city's early leaders which complemented and extended the city's favourable location at the lower end of Lake Michigan, with its excellent harbour and position as crossroad for railways in the midwestern part of the United States. One is well known, and any visitor to Chicago cannot fail to be impressed by it. This was the decision to preserve most of the city's twenty miles of shoreland along Lake Michigan for public use, the vast majority of it in parkland. Given Chicago's history of political corruption and the strong penchant for land speculation pervading the American frontier mentality, it is remarkable indeed that so much of Chicago's shoreland, more than that of any other American city, still remains in public use.

The other decision is much less known. It was made in 1889 when voters in a surrounding 120 square mile area elected to be annexed to Chicago, thus quadrupling the land area of the city overnight (Mayer and Wade 1969). Contrary to the prevailing logic today, residents of the adjacent suburbs then found the opportunities opened up by co-operation with Chicago more attractive than continued independence. With several other annexations over the next twenty-five years, the city reached its present extensive size of 227 square miles by 1915. Because of these territorial decisions, it was able to outpace many of its competitors. Chicago grew rapidly and was able to capture the wealth of sparsely built-up areas within its boundaries as development moved inexorably outwards from its centre. By the end of the nineteenth century it was already the second largest city in the country after New York. By 1950, it reached its population zenith at 3.6 million people, still retaining a reputation as the country's industrial heartland and transportation hub.

In the post-industrial era, Chicago still is seen as one of the world's major cities. Its downtown, the Loop, has become a vibrant centre for advanced corporate services, international trade and tourism. The city still retains a strong industrial base, albeit reduced in size and changed in character. Its airport has more passengers on a yearly basis than any other in the world. But in other respects, Chicago is struggling mightily. Its future is uncertain as it tries, like the proverbial Dutch boy with his finger in the dike, to stem the tide of encroaching forces that have been unleashed on it in the last forty years.

In the lead article of a recent newspaper series on migration patterns within the Chicago metropolitan area, one reporter characterised Chicago as becoming a second-class citizen in the metropolitan region it used to dominate (Reardon 1993).

The fact is that in 1990 Chicago had only a little over a third of the metropolitan area's population, whereas forty years before it had two-thirds. In comparison to the 262 suburban governments in the six-county metropolitan area,[1] Chicago had only a third as many factories, a per capita income nearly 50 per cent lower, 50 per cent fewer jobs, and a poverty rate 4 times higher. (See Figure 4.1 map of the Chicago metropolitan area.)

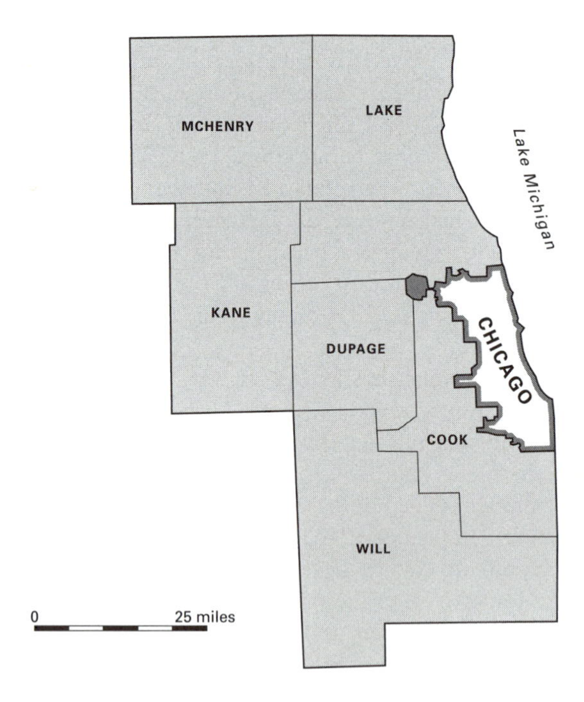

Figure 4.1 The Chicago metropolitan area: city of Chicago and county governments

Almost two-thirds of Chicago's current population are members of minority groups. The largest is the African-American group, which, although comprising 38 per cent of the city's population, decreased in number for the first time in the city's history in the decade of the eighties. Hispanics, who constitute another 19 per cent of the city's population, and Asians, who make up 4 per cent, however, are steadily increasing in numbers. In contrast, the suburban ring population is very much White: 83 per cent of the metropolitan population living outside the city of Chicago are Whites. Despite an increase in the number of minorities who moved to the suburbs from Chicago in the 1980s, the racial divide between city and suburbs is still very wide. And by virtue of the city's continued loss of White middle-income people – the city's population fell from a peak population of 3.6 million in 1950 to 2.7 million in 1990 – the once vaunted political influence Chicago wielded in its state capital in Springfield, Illinois, and even in the national capitol in Washington, DC, has significantly diminished. Chicago is clearly losing political power to its suburbs.

Within the city itself, there are equally troubling signs. Chicago is a highly segregated city, one of the most segregated in the country. In one of every five of its seventy-seven community areas, at least a quarter of the property has been abandoned and now lies vacant, with at least 10 per cent of the total housing units in these areas vanishing during the decade of the 1980s (Chicago Rehab Network 1993). Only 5 per cent of the city's total housing stock is social or subsidised housing (Chicago Department of Housing 1994).[2] This means that the cost of rental housing is higher than it should be for many. For example, 35 per cent of Chicago's renters pay more than 35 per cent of their incomes for rent (Chicago Rehab Network 1993).

Chicago job losses have been immense. From 1963 to 1982, the city lost 269,000 manufacturing jobs, 64,000 retail jobs and 47,000 wholesale jobs, while gaining only 57,000 selected service jobs. This amounted to a net loss of 323,000 jobs for that period (Wilson 1994). Job growth projections to the year 2010 for the city are modest at 12 per cent. Much of the estimated expansion is expected to occur in the FIRE sector centred in the city's downtown, but this would be balanced by continued sizeable manufacturing job losses in the rest of the city (Northeastern Illinois Planning Commission 1988).

The situation of Chicago's Black inner city 'is emblematic of the social changes that have sown despair and exclusion in the ghettos of Northern metropolises . . . an unprecedented tangle of social woes is now gripping the Black communities of Chicago's South Side and West Side' (Wacquant and Wilson 1989: 11). That growing poor population is an extraordinarily difficult conundrum confronting the city. Despite the city's recent success in being designated one of six empowerment zone cities by the Clinton White House, making it eligible for $100 million in grants spread over ten years – an acknowledgement that the city still has some clout within the Democrat-controlled White House – it is doubtful that any more meaningful assistance will be forthcoming from the federal government for cities like Chicago in the coming years to help lessen their burden of sheltering an increasingly dependent poor population. Undoubtedly, the November 1994 elections, with conservative Republicans sweeping to power in the US Congress and State capitals, will push the priorities of older communities like Chicago even further down the list, making matters even worse for them.

In this chapter, I will look more closely at the issue of spatial segregation in Chicago, and the social exclusion and social polarisation that accompany it. What is the extent of segregation in Chicago, what are some of the reasons for it, and what are its effects?

The extent of segregation in Chicago and some reasons for it

Chicago has the dubious distinction of being, if not the most segregated city in the United States, among the most segregated. Massey and Denton (1993) give the fullest picture of the extreme pattern of segregation of Chicago's African-American population. Using data from the 1980 US census, they present five measures of Black segregation in the thirty US metropolitan areas with the largest Black populations. These are used to derive a list of what they call the most hypersegregated areas of the United States (see Table 4.1).[3] These five dimensions of segregation include measures to gauge the extent of unevenness (Blacks being over-represented in some areas and under-represented in others), isolation (Blacks sharing or not sharing a neighbourhood with Whites), clustered (Black neighbourhoods tightly clustered together or scattered about in checkerboard fashion), concentrated (Black neighbourhoods concentrated within a very small area or settled sparsely throughout the urban environment), and centralised (Black neighbourhoods centralised around the urban core or spread out along the periphery).

The Chicago area is clearly hypersegregated. It ranks high on every one of the five indices, on three of them ranking either first or second highest.[4] Based on its rank on each of the five measures, it has the highest overall score of any of the thirty metropolitan areas. *It thus represents the most hypersegregated metropolitan area in the United States.*

Massey and Denton updated one of the five indices for the Chicago metropolitan area to 1990, the index of unevenness or Black–White residential dissimilarity.[5] They found that it had changed very little from 1970 to 1990. In 1970, the index stood at 92. By 1980, it had dropped to 88 and in 1990 it dropped slightly again to 86, within 6 points of its 1970 value (p. 222). As a point of comparison, it is interesting to note that the same segregation index computed for Turks, Moroccans and Surinamese living in Amsterdam ranged from 34 to 36 depending on the group (Musterd and Ostendorf 1994).

A closer look at Chicago's community areas confirms the extensiveness of segregation in the city. (Figure 4.2 shows the location of community areas in Chicago in 1990 with 50 per cent or more African-American and Hispanic

Table 4.1 Five dimensions of Black segregation in the thirty US metropolitan areas with the largest Black population, 1980

	Dimension of segregation				
Metropolitan area	*Unevenness*	*Isolation*	*Clustering*	*Centralisation*	*Concentration*
Northern areas					
Boston	77.6	55.1	49.1	87.1	79.9
Buffalo	79.4	63.5	44.5	88.4	88.2
Chicago	*87.8*	*82.8*	*79.3*	*87.2*	*88.7*
Cincinnati	72.3	54.3	15.8	88.3	66.9
Cleveland	87.5	80.4	74.3	89.8	92.7
Columbus	71.4	57.5	32.1	93.3	85.4
Detroit	86.7	77.3	84.6	92.4	84.2
Gary–Hammond–E. Chicago	90.6	77.3	56.1	88.7	86.9
Indianapolis	76.2	62.3	41.1	94.2	85.7
Kansas City	78.9	69.0	46.1	92.1	85.7
Los Angeles–Long Beach	81.1	60.4	76.5	85.9	69.5
Milwaukee	83.9	69.5	68.9	95.1	94.4
New York	81.6	62.7	46.8	79.5	89.2
Newark	82.0	69.2	75.5	85.9	91.9
Philadelphia	78.8	69.6	67.3	85.5	75.7
Pittsburgh	72.7	54.1	27.2	81.2	82.1
St Louis	81.3	72.9	26.4	93.1	89.3
San Francisco–Oakland	71.7	51.1	28.2	83.6	68.7
Average	80.1	66.1	52.2	88.4	83.3
Southern areas					
Atlanta	78.5	74.8	39.8	82.7	68.6
Baltimore	74.7	72.3	62.2	85.7	76.3
Birmingham	40.8	50.2	5.9	83.0	77.5
Dallas–Ft Worth	77.1	64.0	33.4	74.9	69.3
Greensboro–Winston Salem	56.0	50.1	5.3	60.1	61.3
Houston	69.5	59.3	23.8	84.0	56.9
Memphis	71.6	75.9	44.0	81.7	55.0
Miami	77.8	64.2	34.4	46.3	56.5
New Orleans	68.3	68.8	32.7	90.6	58.4
Norfolk–Virginia Beach	63.1	62.8	19.9	71.2	55.9
Tampa–St Petersburg	72.6	51.5	24.6	58.1	49.3
Washington, DC	70.0	68.0	45.0	85.0	44.1
Average	68.3	63.5	30.9	75.3	60.8
'Hypersegregation' cutpoint	60+	60+	60+	60+	60+

Source: Douglas S. Massey and Nancy A. Denton (1989) 'Hypersegregation in U.S. metropolitan areas: Black and Hispanic segregation along five dimensions', *Demography* 26: 378–9

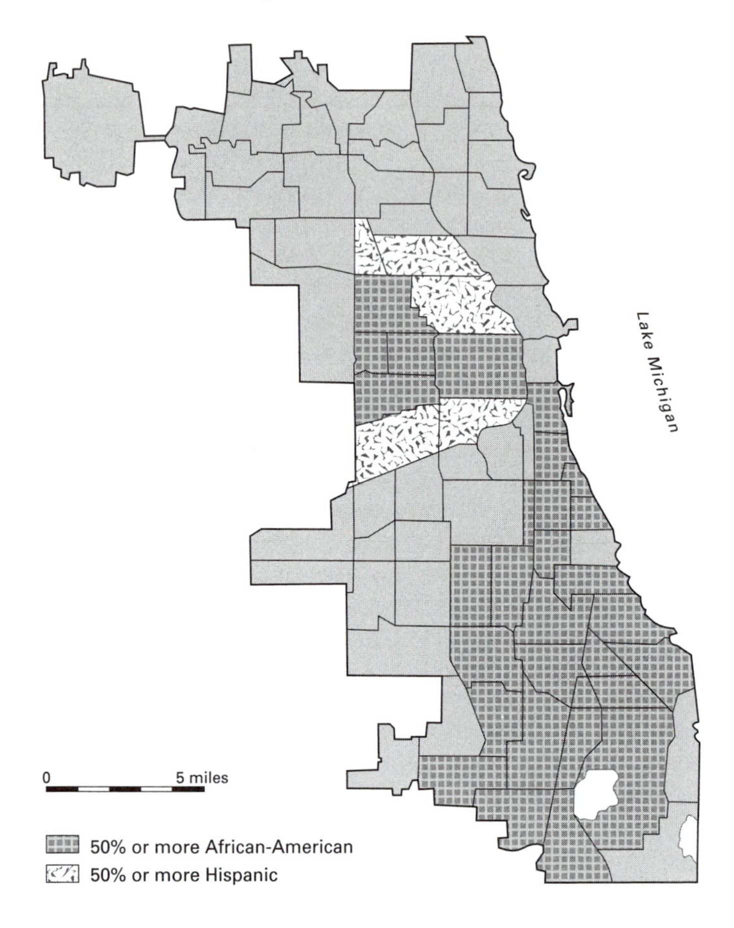

Figure 4.2 Chicago community areas with more than 50 per cent African-American and with more than 50 per cent Hispanic population, 1990

Source: City of Chicago Comprehensive Housing Affordability Strategy report, 1994

people.) In 1990, almost a third (31 per cent) of the city's 77 community areas, or 24 areas, had a concentration of African-Americans exceeding 85 per cent. Of these 24 areas, 4 of every 5 of them were virtually all Black, each having 96 per cent or more Blacks. And all but one of these most segregated community areas had virtually the same percentage of Blacks in 1980.

Looking at where White people lived in the city in 1990, 15 community areas had at least 85 per cent or more Whites.[6] All of these areas were on the periphery of the city, half in the Northwest and the other half, except for one, in the Southwest (Chicago Department of Planning and Development 1993). If the most segregated White and Black areas (the ones having more than 85 per cent of their respective populations in 1990) were added together, then

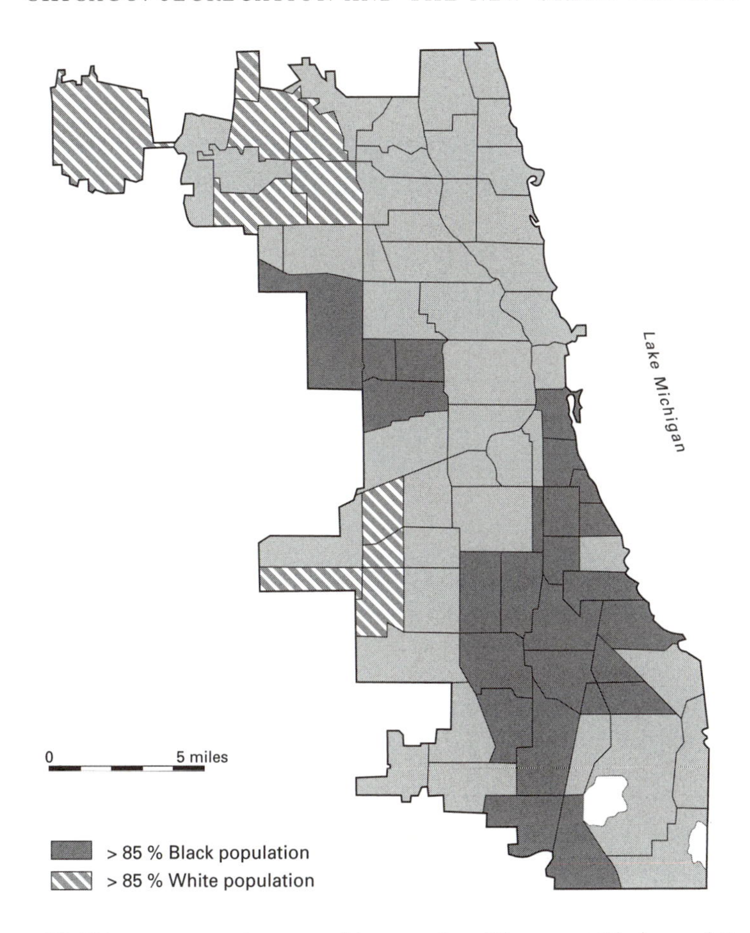

Figure 4.3 Chicago community areas with more than 85 per cent Black population and with more than 85 per cent White population

Source: Chicago Department of Planning and Development, *Population by Race*, 1993

fully half of Chicago's 77 community areas could be considered extremely segregated by race (see Figure 4.3).

In contrast to the concentrated African-American and White areas, the Hispanic population, which is half the size of the African-American population, appears to be spread out more. In only two community areas did Hispanics comprise more than 80 per cent of the population; in 26 other areas they comprised from 10 to 69 per cent of the population.

It appears that African-Americans in Chicago are highly segregated from Whites even at higher-income levels. The average Black–White dissimilarity index in 1980 for wealthier African-Americans in the Chicago area earning more than $50,000 a year was 86. This is only 5 points less than the comparable figure for poor African-Americans earning under $2,500 a year (Massey and Denton 1993: 86). Black segregation appears to be universally high, impervious to income differentials, while that of Hispanics and Asians 'falls progressively as status rises' (Massey and Denton 1993: 88).

As African-Americans move out of Chicago to the suburbs, a similar pattern of spatial separation from Whites appears to be taking place. In 1980, 16 per cent of African-Americans living in the Chicago metropolitan area made their homes in the suburbs. By 1990 that percentage had risen to 24 per cent.[7] The number of suburbs where Blacks made up more than 50 per cent of the population increased from seven to thirteen in the 1980s. And in virtually every one of the suburbs that experienced a large increase of Blacks and other minorities, there was also a decline in the White population. But the most telling statistic indicating that the racial separation pattern of the city of Chicago is being replicated in the suburbs is that four of every five new Black suburbanites in the 1980s (81,000 of 103,000 people) resided in only 8 per cent of the region's 262 suburbs, or 22 suburbs (Fegelman 1991). Most of these suburbs were in southern Cook County which abuts the Black belt on Chicago's south side.[8]

Different reasons are given for why a city like Chicago is so segregated. Racism is clearly one of them – i.e. a complex system of institutional racism keeps Blacks penned up in smaller, concentrated areas. The structural transformation of the economy is another explanation, the argument being that large job losses in the higher-paying manufacturing sector resulted in intensifying the pattern of poor people concentrated in ghettos.

Governmental actions are also seen as contributing forces. Many believe that urban renewal and expressway projects in Chicago destroyed some viable lower-income neighbourhoods where Blacks lived and the supply of affordable housing units in them. Not only was the noose tightened around the remaining low-income areas, but a new vertical ghetto was built, high-rise public housing projects along one of the city's major expressways, the Dan Ryan expressway, to rehouse the displaced poor from 'renewed' areas (Wacquant 1989).

Policies of banks and insurance companies to redline Black neighbourhoods on the one hand, thus restricting the flow of mortgage and insurance loan money to them, and on the other hand to deny loans to some Blacks who wanted to move out of the ghetto, were also considered factors that limited the spatial mobility of Blacks and reinforced the segregation pattern.[9]

The fragmented governmental system in the metropolitan area is also seen as facilitating segregation. Because suburban governments rely so heavily on the property tax for revenue, they often compete for developments that bring in more tax revenues than cost in expenditures. Thus, just on rational fiscal grounds (leaving aside the race factor), most suburban governments would frown upon building affordable housing developments to serve lower-income families as too costly to them. Few poor people from Chicago therefore move out to the suburbs and inner-city ghettos remain.[10]

The effects of segregation in Chicago

Some of these explanations work together to perpetuate the pattern of segregation. Rather than dwell more on the reasons for the extreme concentration of Black population in the city, let me discuss some of the effects such segregation is having, particularly on those areas with heavier concentrations of poor people.

Two leading proponents of the view that minority residents of large industrial metropolises have been especially hurt by the accelerating structural economic changes of the past several decades contend that

> The shift to service-producing industries, together with the relocation of plants abroad or to cheaper labour sites nationally (in the suburbs and in the South, where unions are weak and employers find themselves in a buyer's market), has led to enormous declines in entry-level blue-collar jobs, particularly in the older central cities of the Middle-Atlantic and Midwestern regions. . . . Because of the disproportionate concentration of Blacks in these good-producing industries, particularly heavy manufacturing, such massive job cutbacks have disproportionately affected central-city Blacks, and the poor in particular.
>
> (Wacquant and Wilson 1989: 79)

The belief of people like William Julius Wilson and his colleagues at the University of Chicago's Center for the Study of Urban Inequality is that the social and institutional structure of older Black ghettos was transformed. Increased joblessness, poverty and receipt of welfare were the consequences, along with an exodus of working- and middle-class families from these areas. Wacquant and Wilson (1989) show that steep rises occurred in several selected social and demographic characteristics for Chicago's ten poorest inner-city neighbourhoods between 1970 and 1980 (see Table 4.2). Poorer residents of these areas, they contend, were put 'in a radically more constraining situation than their counterparts of earlier times' (Wacquant and Wilson 1989: 95).

This new urban poverty is characterised by segregated neighbourhoods inhabited by poor Blacks in which a substantial majority of individual adults are either unemployed or have dropped out of the labour force. Wilson (1994) looked at fifteen overwhelmingly Black and poor community areas in Chicago to illustrate this phenomenon. Forty per cent of the city's Black population, 425,125 people, lived in these areas in 1990. The average poverty rate in twelve of them exceeded 40 per cent, with the poverty rate ranging from 29 to 36 per cent in the other three. He points out that only one in three adults held a job in the twelve poorest areas, while the comparable figure for the other three areas was four in ten. So an average of only 37 per cent of all the adults in these fifteen areas were gainfully employed. By comparison, the city-wide average was 57 per cent, only a few percentage points higher than what

Table 4.2 Selected social and demographic characteristics of Chicago's ten poorest inner-city neighbourhoods, 1970–80

	% of families below poverty level		*% of families headed by a female*		*% of adults (aged 16 or over) not in labour force*		*% of population on AFDC-GA*		*% change, 1970–80*		
	1970	*1980*	*1970*	*1980*	*1970*	*1980*	*1970*	*1980*	*Population*	*Net migration*	*Number of AFDC-GA recipients*
South Side											
Near South Side	37.2	42.7	41.0	76.0	55.2	62.4	22.4	72.8	–17.4	–28.0	+168.6
Douglas	31.1	42.6	43.0	70.0	48.9	57.0	24.3	36.6	–13.5	–33.0	+30.6
Oakland	44.4	60.9	48.0	79.0	64.3	76.0	38.4	60.5	–8.4	–25.6	+44.1
Grand Boulevard	37.4	51.4	40.0	76.0	58.2	74.5	30.4	45.6	–32.9	–37.6	+0.7
Washington Park	28.2	43.2	35.0	70.0	52.0	67.1	23.2	48.2	–30.6	–35.7	+50.5
Englewood	24.3	35.8	30.0	57.0	47.7	61.9	21.8	41.4	–34.2	–46.2	+25.0
West Side											
Near West Side	34.7	48.9	37.0	66.0	44.6	64.8	26.9	44.4	–27.2	–37.9	+20.1
East Garfield Park	32.4	40.3	34.0	61.0	51.9	67.2	32.5	42.7	–39.5	–53.8	–20.6
West Garfield Park	24.5	37.2	29.0	58.0	47.7	58.4	24.6	40.4	–30.1	–46.7	+14.8
North Lawndale	30.0	39.9	33.0	61.0	56.0	62.2	32.2	40.6	–35.1	–50.1	–18.0
Chicago	12.2	16.8	29.7	27.0	41.5	44.8	8.5	16.9	–10.8	–17.3	+78.1

Sources: Chicago Fact Book Consortium (1984), City of Chicago (1973) and Chicago Area Geographic Information Study (no date)

the average was in the seventeen other predominantly Black but relatively higher-income community areas in Chicago.

Wilson sees joblessness coupling with the decline of social organisation in these concentrated poverty areas – weaker social networks and much less collective supervision by the residents themselves in addressing neighbourhood problems – as the critical factors leading to more crime, gang violence, drug trafficking, family break-ups and problems in the organisation of family life.

Massey and Denton (1993) start where Wilson leaves off. They acknowledge that the structural transformation of the economy harmed many racial and ethnic groups, as manufacturing jobs disappeared and a two-tier service economy job sector arose. They also agree that the economic structural transformation played a crucial role in creating the urban underclass, with its much greater degree of joblessness.[11]

But they go further than Wilson by contending that what made the underclass a disproportionately Black underclass was racial segregation. Black Americans, they say, were the most highly segregated minority group, so among them, more than other minority groups, the resulting income loss from the economic restructuring was confined to a small set of spatially contiguous and racially homogeneous neighbourhoods.[12] And the concentration of poverty due to segregation intensified the problems for Blacks living in such areas.

They attempt to show this in two ways. The first is more conceptual. They devise a hypothetical experiment depicting several ideal cities where key features are held constant except for the level of racial segregation. The second is by comparing the actual effects that factory closings on Chicago's West Side had on two West Side neighbourhoods adjacent to each other, an African-American and a Mexican-American neighbourhood.

In their experiment they create four hypothetical cities, each having the same population size (126,000 people), the same overall racial composition (25 per cent Black and 75 per cent White), the same number of neighbourhoods (16), and the same Black and White poverty rates (20 per cent for Blacks and 10 per cent for Whites). The only thing varied is the level of racial segregation in the city. On one extreme is a city with no racial segregation (a Black–White segregation score of 0 per cent). Thus each of the sixteen neighbourhoods in this hypothetical city would have 2,000 Blacks and 6,000 Whites. At the other extreme is a city with complete racial segregation (a 100 per cent Black–White segregation score). This translates into four all-Black neighbourhoods and twelve all-White neighbourhoods. Between these they postulate a city with low racial segregation – a 33 per cent Black–White segregation score – and another one with high racial segregation – a 66 per cent Black–White segregation score.

As part of their analysis, they devise a simulation to show the effects of an increase in the Black poverty rate from 20 to 30 per cent (the result of economic restructuring) in both of the polar opposite cities – the one with no racial segregation and the other with complete racial segregation. Table 4.3 shows the results of their analysis for a series of neighbourhood conditions – household

Table 4.3 Simulated effects of a shift in the Black poverty rate on neighbourhood conditions experienced by poor Blacks assuming different levels of racial segregation

Neighbourhood condition and level of racial segregation	*Effect of shift in Black poverty rate*		
	Poverty rate of 20%	*Poverty rate of 30%*	*Change*
% of houses boarded up			
No racial segregation	1.2	1.5	0.3
Complete racial segregation	2.3	3.3	1.0
Major household income ($)			
No racial segregation	13,020	11,235	1,785
Complete racial segregation	8,160	4,523	3,637
% of families on welfare			
No racial segregation	21.1	24.6	3.5
Complete racial segregation	36.1	51.0	14.9
% of female-headed families			
No racial segregation	19.2	22.2	3.0
Complete racial segregation	33.5	45.5	12.0
% of high school students scoring below 15th percentile			
No racial segregation	32.6	35.3	2.7
Complete racial segregation	47.1	57.8	10.7

Source: Douglas S. Massey (1990) 'American apartheid: segregation and the making of the underclass', *American Journal of Sociology* 96: 343, 348–9

income, boarded-up houses, crime, welfare families, female-headed families, and high school student performance.

When poor Blacks live in a city with no racial segregation (one in which Whites with higher incomes are always living in the same neighbourhood as the Blacks), the neighbourhood has more overall disposable resident income to draw upon for housing improvements, neighbourhood businesses and other general neighbourhood improvements. Though overall neighbourhood income declines when the Black poverty rate rises in the non-segregated city, the decline is modest and the negative ripple effects on neighbourhood conditions are not seen as severe.

But when poor Blacks live in completely segregated areas, and, furthermore, when the poverty level rises in such areas, Massey and Denton posit many more serious difficulties. 'When the Black poverty rate rises in a totally segregated city, the drop in income and potential demand is confined to neighbourhoods inhabited exclusively by Blacks, and primarily by poor Blacks, leaving the latter trapped in neighbourhoods with insufficient income' (1993: 136). Thus, for example, there is less money to pay for home improvements and sustain a viable retail sector. In addition, other negatives occur more frequently – e.g., more crime, growing family disruptions, increasing educational failure. In summary, they argue that a rise in poverty under conditions of high racial segregation leads to a concentration of poverty for the segregated group which then sets off a series of negative changes in the social and economic conditions of such neighbourhoods.

Massey and Denton ground their hypothetical simulation with a concrete example from Chicago. Drawing on information from a Chicago Tribune publication about underclass areas in Chicago (Chicago Tribune 1985), they contrast two neighbourhoods separated by only a few hundred metres, one inhabited predominantly by African-Americans and the other by Mexican-Americans. Both neighbourhoods felt the economic shocks of factory closings in the 1970s on Chicago's West Side. But they claim that the economic transformation hit the African-American neighbourhood, North Lawndale, much harder than it did the Mexican-American neighbourhood, Little Village.

Two huge factories anchoring the North Lawndale neighbourhood, and the world headquarters of Sears Roebuck located in its midst, provided low-skilled jobs and steady income for many West Side residents. These big employers either left or gradually phased down their operations from the end of the 1960s and through the 1970s. As these big plants left, so did numerous smaller stores, banks and many other businesses in North Lawndale dependent for their sales on the wages these large employers had paid their workers who were laid off. It was estimated that the community lost 75 per cent of its business establishments from 1960 to 1970. Today North Lawndale has 48 state lottery agents, 50 currency exchanges and 99 licensed bars and liquor stores, but only one bank and one supermarket for a population of some 50,000. Added to this was the disappearance of nearly half of North Lawndale's housing stock since 1960, a soaring crime rate, a sharply increased infant mortality rate and a sharp increase in the percentage of those receiving welfare assistance.

While North Lawndale began to resemble a war zone, Massey and Denton say that despite the economic recession the close-by Mexican-American neighbourhood of Little Village remained a hive of commercial activity throughout the 1970s and 1980s. The Little Village shopping district continued to house a variety of supermarkets, banks, restaurants, bakeries, travel agencies, butchers, auto shops and other retail stores. They say that the key difference between the two areas affected by the wave of factory closings on the city's West Side was the different degree of segregation each experienced. Little Village, although predominantly Mexican-American in 1980, still had about a 25 per cent non-Hispanic population. North Lawndale, in contrast, was only 2 per cent non-Black.

> Because of this disparity in the degree of segregation, the economic dislocations of the 1970s brought an acute withdrawal of income from North Lawndale, pushing it well beyond the threshold of stability into disinvestment, abandonment, and commercial decline; but the same economic troubles brought only a moderate concentration of poverty in Little Village, leaving it well shy of the tipping point.
>
> (Massey and Denton 1993: 137)

William Julius Wilson, in the work that his Center for the Study of Urban Inequality has been doing on a larger study of poverty in Chicago, appears to have come round to the position that racial segregation matters more than he acknowledged in his previous books (1978 and 1987). Some criticised him for downplaying the race factor in his earlier works. Wilson now, however, says clearly that 'If large segments of the African American population had not been historically segregated in inner city ghettos we would not be talking about the new urban poverty' (1994: 10).

He acknowledges that Massey and Denton are correct in saying that the segregated ghetto is the product of systematic racial practices – e.g. redlining by banks and insurance companies, panic peddling by real estate agents, the creation of massive public housing projects in low-income areas, the effects of government actions like urban renewal and freeway building and restricted Black residential mobility.

But pointing out that Black segregated neighbourhoods existed then and now, Wilson says: 'Given the existence of segregation, one then has to account for the ways in which other changes in society interact with segregation to produce the recent escalating rates of joblessness and problems of social organisation' (1994: 11). My reading of Massey and Denton leads me to conclude that they too give credence to the economic structural transformation factor as a key force in contributing to lower income, higher poverty rates and more joblessness for Blacks. One might conclude that they differ in emphasis on what's the more important explanation for the new urban poverty. Wilson accentuates more strongly the economic structural reasons than the racial segregation reasons as causing the new urban poverty while Massey and Denton place more emphasis on racial segregation as the more important explanatory factor. So the differences between these important American scholars seems to be fading on the analytical side although their current works still show divergence in the main thrust of their policy prescriptions.

Conclusions

Chicago represents an advanced case of a segregated city. There is no counterpart on the European continent. To some extent, its segregation pattern has been influenced by the US welfare state model. There would probably be less segregation in Chicago if the USA had a welfare state system more like Sweden, the Netherlands, or France. For one thing, more social housing would undoubtedly have been available in Chicago than the minuscule 5 per cent of the city's housing stock now existing. The more social housing there was, the more likely more working-class people of different races might be living in the same neighbourhoods as lower-income Blacks.

A policy of full employment at the national level, with effective job training and public service as well as private sector jobs that would reach the growing number of jobless people in Chicago, would, as incomes rise, also have helped to reduce the level of spatial segregation. Social insurance programmes that were not rigidly means tested, which could have reduced the per capita cost of health care and possibly even provided a guaranteed minimum income, could also have reduced the spread of segregation.

But such programmes are obviously not on the cards. It would be sheer fantasy to think that in the late twentieth century the United States might move in the direction of where the European social democratic or corporatist state welfare countries are, as Esping-Andersen (1990) classifies them, even though many of these countries are apparently moving in the direction of the US liberal welfare state system.

But even if the USA were suddenly transformed into a welfare state more closely approximating that of a European social democratic or corporatist country, would racial segregation in the housing sphere and the social exclusion that accompanies it, as it exists today in Chicago, be drastically reduced? I think not. The explanation lies not so much with the changes in welfare state policy, but with the uniquely American and seemingly intractable dilemma that Gunnar Myrdal wrote so perceptively about fifty years ago in his classic study of the American racial problem, *An American Dilemma* (Myrdal 1944).

While many changes have occurred since then to soften the edges of racial prejudice and discrimination in the job market, in schools, and in community policies as a result of judicial decisions, laws and administrative regulations, discrimination in the housing market on the basis of race is still deeply ingrained in American society. And this, in combination with the economic structural changes that have occurred, leads to sustaining and even extending hypersegregation (Massey and Denton 1993) and hyperghettoisation (Wilson 1994) in the late twentieth century in cities like Chicago.

Spatial segregation of poor people, especially African-Americans in Chicago, is extreme and there is strong evidence to support the contention that it has compounded considerably their chances of moving up the social mobility ladder. The prospects for many of that population today are exceedingly limited.

Chicago has always been a city of immigrants. By 1890 nearly 78 per cent of the city's population was foreign-born or children of the foreign-born (Mayer and Wade 1969). The early settlers were English, Germans, Scandinavians and Irish. Towards the end of the nineteenth century, Poles, Czechs, Italians and Eastern European Jews streamed into the city in large numbers. Most of these groups settled first in older and congested housing in cramped areas close to the centre of the city. They followed a similar pattern once they got more income and became more established, moving further away from the centre of the city to roomier and more pleasant quarters.

Chicago had a small settlement of Blacks as far back as the 1860s, with their numbers increasing slowly as the demand for labour in the city's stockyards and other heavy industries intensified. Although still only 2 per cent of the city's population in 1910, the Black population started to grow more rapidly. By the early 1920s, a long, narrow band of Black occupancy stretched south from the central area of the city, in addition to other smaller concentrations on the north and west side. (See Figure 4.4, which shows the areas with 25 per cent or more Negro population that were added from 1920 through 1965.) But the mobility pattern for Blacks in the early twentieth century was different than that for other immigrant groups. Mayer and Wade (1969: 252, 254) graphically describe it:

> Everywhere the Negro moved, the grim spectre of segregation was sure to follow. . . . Negro housing bore the marks of this discrimination. . . . Few Negroes owned their own homes. . . . Rents for Negro housing

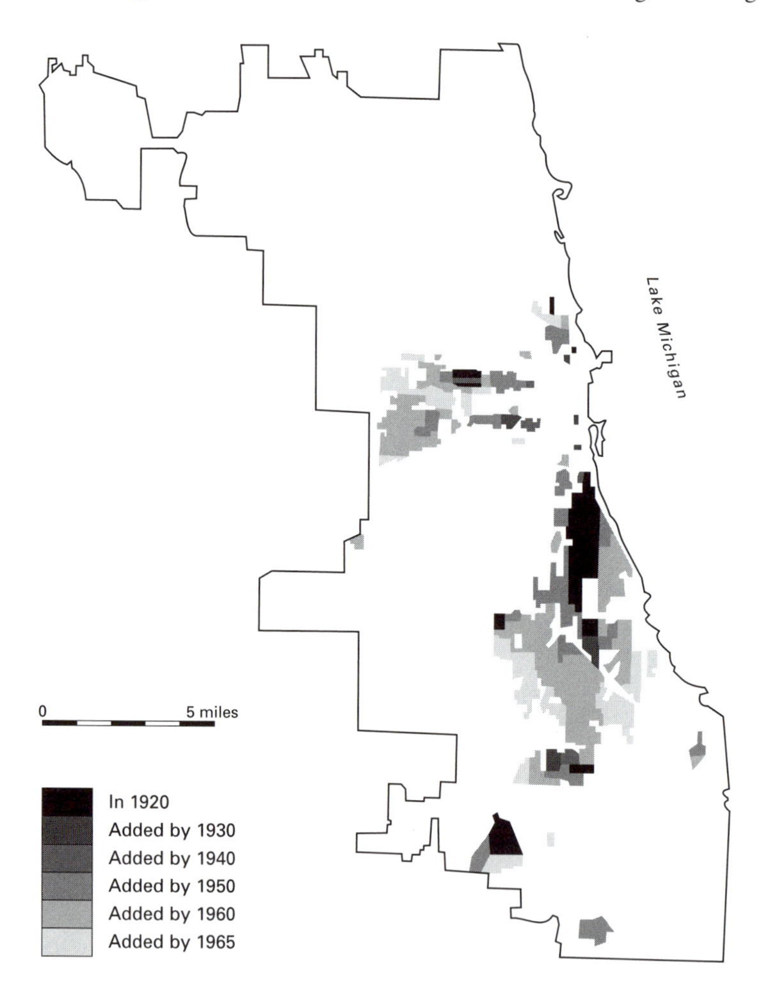

Figure 4.4 Areas with 25 per cent or more Negro, 1920–65
Source: Mayer and Wade (1969: 411)

> were never less than for White housing and commonly ran twenty-five per cent more. . . . The Negro was not given the same options as the foreign immigrants from Europe even if he managed to accumulate wealth and education; his world remained the ghetto with its small comforts and little hope.

During the Depression years of the 1930s, Black families in Chicago, particularly those on relief, lived tightly packed together in a small geographical area mostly on the city's south side (Drake and Cayton 1945). The Black ghetto steadily increased in size in Chicago as a result of the economic opportunities northern cities afforded southern Blacks. Many trekked north in the late 1930s, 1940s and 1950s from the farms they worked on as sharecroppers, because southern agricultural production became mechanised and therefore less labour-intensive. The prospect of finding jobs in the factories of big northern cities like Chicago lured great numbers of them to move north. Ironically, many of their offspring now find themselves adrift in cities like Chicago as a result of the economic structural transformation that has taken place in the past few decades (Lehmann 1991).

Today, the African-American population comprises nearly 40 per cent of the city's population and Blacks live in an ever-widening belt on the city's South Side and West Side (see Table 4.2). More and more Blacks have moved up the social mobility ladder, living in more comfortable dwellings and pleasant neighbourhoods. Others, also increasing in number, however, live in zones of squalor and fear, under conditions that most Americans would abhor if they had to live there. Conditions in parts of the greatly expanded Black ghetto are bad and getting worse, triggered by the economic transformation and the debilitating effects of having to live in segregated, concentrated poverty areas.

What happens to the new poverty population clearly has a bearing on the capacity of the city of Chicago to regenerate itself for the twenty-first century. But the albatross of this increasingly dependent and destitute population may prevent the city from reaching its fuller potential. Clearly, Chicago, by itself, cannot solve the problem of its new poor population. The problem is regional, national and global in scale and scope. Answers to the problem exist, but the capacity to carry out what is needed seems more and more remote.

Notes

1 The region is very fragmented. All told there are 1,200 separate governments in the six-county Chicago metropolitan area including the suburban governments, township governments, elected school boards and a host of other special district governments.

2 The 5 per cent social or government-subsidised housing amounts to approximately 60,000 units. Of these, 40,000 are public housing units administered by the Chicago Housing Authority, 10,000 of which are for elderly people only. In fact, no family public housing developments have been built by the Chicago Housing Authority

since 1969. Since 15 per cent of the authority's public housing units are currently vacant, the total number of occupied public housing units is actually only 34,000, yet in 1990 the waiting list for public housing was more than 26,000 households. The other 20,000 subsidised housing units include developments built and managed by private developers as well as Section 8 rental subsidy units managed by the Chicago Housing Authority (Chicago Department of Housing 1994).

3 Any index that exceeds a value of 60 is considered to constitute 'high' segregation. Metropolitan areas that display high segregation on at least four of the five dimensions are considered to be hypersegregated.

4 Computing the indices for unevenness and isolation for just the city of Chicago, not for the entire metropolitan area, shows that Chicago ranked first among the thirty cities in the Massey–Denton sample on both these measures in 1980 (see Table 3.3, p. 71).

5 This index measures the percentage of all Blacks who would have to move to achieve an even or 'integrated' residential configuration – one where each census tract replicates the metropolitan area as a whole.

6 In 1980 there were 21 community areas with a concentration of Whites exceeding 85 per cent. The decline by 6 in the number of highly concentrated White community areas from 1980 to 1990, to 15, is most likely because of the sizeable out-migration of Whites to the suburbs during the 1980s decade – there were approximately 250,000 fewer Whites in Chicago in 1990 than in 1980, a 19 per cent drop in the city's White population.

7 In 1990, 35 per cent of all Hispanics in the metropolitan area were in the suburbs. For Asians, the figure was considerably higher – almost 60 per cent were residing in the suburbs by 1990. This suggests a much faster rate of migration from city to suburb among Asians than other minority groups, although some more educated and middle-income Asians probably move directly to the suburbs rather than into Chicago when they first arrive in the region.

8 In contrast, suburban Hispanics and suburban Asians were much more dispersed in the region than were suburban Blacks. Data compiled by the author from a regional planning agency report (Northeastern Illinois Planning Commission 1991) show that in 1990, 90 per cent of all suburban Hispanics resided in 99 suburbs and 90 per cent of all suburban Asians resided in 87 suburbs. Ninety per cent of the suburban Blacks, however, resided in only 35 suburbs in 1990.

9 The paucity of loans from financial lending institutions to poor areas of Chicago is quite evident today: 33,692 housing loans were made by lending institutions in Chicago in 1991, amounting to $2.9 billion. Of this amount, only 1 per cent ($32.7 million) reached the seven poorest low-income African-American communities on the south and west sides of the city (Chicago Rehab Network 1993).

10 The one exception in the Chicago area is the well-known Gatreaux judicial decision. The case involved a class action suit filed in 1966 in federal court charging the Chicago Housing Authority and the US Department of Housing and Urban Development with discrimination in locating federally assisted housing projects in all Black neighbourhoods of the city. After a long and protracted battle, an agreement was approved in 1981 to promote the desegregation of public housing in the Chicago area. This required the Chicago Housing Authority to grant rent subsidy vouchers to 7,100 Black families over the next ten years. Many of these vouchers were used by poorer Blacks to move to housing developments already built in some of the suburbs (Massey and Denton 1993: 190–1).

11 It should be noted that although Wilson's and Massey and Denton's works are mostly cited in this chapter, a considerable literature about the underclass population in the United States exists. Among important works by scholars focusing on

different aspects of the underclass issue are ones by Andersen (1990), Gans (1990), Glascow (1980), Jencks (1992), Jencks and Peterson (1991), Kasarda (1989) and Ricketts and Sawhill (1988).

12 Massey and Denton contend that Puerto Ricans are most like Blacks, in the sense that they have simultaneously experienced high levels of residential segregation and sharp increases in poverty in the past couple of decades. Puerto Ricans, they say, are the only Hispanic group whose segregation indices are routinely above 70. They attribute the higher degree of segregation of Puerto Ricans among Hispanic groups to the fact that a large number of them are of African origin, and thus darker-skinned (1993: 146–7).

5

THE WELFARE STATE, ECONOMIC RESTRUCTURING AND IMMIGRANT FLOWS

Impacts on socio-spatial segregation in Greater Toronto

Robert A. Murdie

Toronto: the post-Second World War context

In the decades since the Second World War Toronto has become the largest and financially most important metropolitan area in Canada. It is also one of the most culturally diverse urban centres in the country. Population in the Toronto Census Metropolitan Area (Figure 5.1) more than doubled between 1961 and 1991, from 1,824,500 to 3,893,000.[1] Toronto's rapid growth during the last few decades can be attributed to several factors. Among these are (1) the opening of the St Lawrence Seaway in 1959, making Toronto accessible to ocean-going ships, (2) the signing of the Automotive Trades Agreement with the United States in 1965, which led to the expansion and development of automobile assembly and parts plants in the Toronto area, (3) the migration, beginning in the 1970s, of English-language business and population from Montreal to Toronto as a result of the implementation of French language laws in Quebec and the possibility of Quebec's separation from Canada, and (4) the relative attractiveness of Toronto (and Vancouver) rather than Montreal as a reception area for Canada's immigrant population (Yeates 1991, Marshall 1994).

Toronto has become the most important financial and business centre in Canada. Coffey (1994: 33–4) notes that as a corporate and financial centre Toronto is now more than twice as important as Montreal. In 1986, it had 41 of Canada's top 100 industrial corporations and 51 of the country's top 100 financial institutions. The corresponding figures for Montreal were 18 and 17 respectively (Marshall 1994: 64–5). Toronto also dominates air passenger flows in Canada and is the headquarters for the Globe and Mail's *Report on*

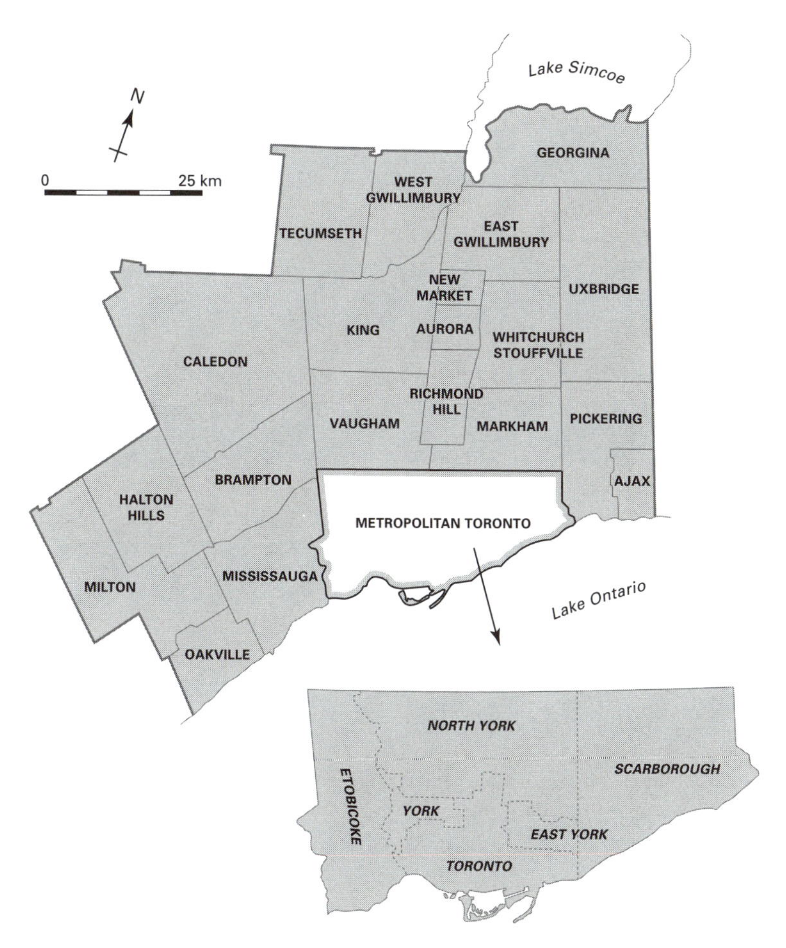

Figure 5.1 The Toronto Census Metropolitan Area

Source: Murdie (1996: 209), Figure 9.1

Business, Canada's major business paper. As Preston (1991: 168) notes, 'The nation's business has become Toronto's business and vice versa.'

Toronto is also the major reception area for Canada's immigrant population and one of the ethnically most diverse metropolitan areas in the country.[2] In each year since 1971 the Toronto area has been the intended destination of more than one-third of Canada's immigrants (Employment and Immigration Canada 1992: 72 and 74, Hulchanski 1993: 23). The number of new immigrants coming to the Toronto area has fluctuated substantially between 1971 and the early 1990s, reaching a low of slightly more than 20,000 in 1983–4 and increasing to about 85,000 in 1992–3 (Metropolitan Toronto Planning Department 1993a: Figure 3.2, 1995: Figure 1.16). More generally, immigration to Toronto increased dramatically in the late 1980s and early 1990s compared to the late 1970s and early 1980s. This is largely due to the higher quotas set by the Canadian government and the attractiveness of Toronto as a destination within Canada. By 1991 about 38 per cent of Toronto's population was born outside Canada and less than half of the population was of British or French origin (1991 Census of Canada). The origins of the immigrant population have also changed dramatically in the post-Second World War period. Of all immigrants living in the Toronto Census Metropolitan Area (CMA) in 1991, 90 per cent of those who came before 1966 were from Europe. In contrast, only 21 per cent of those who arrived in the 1986–91 period originated from Europe. Just over half were from various Asian countries and an additional 25 per cent were from countries in Africa, Central and South America and the Caribbean (Lapointe and Murdie 1995: Table A1 and Figure 5.2 in this chapter).

The life chances of these immigrants, as well as the Canadian-born, are determined to a considerable extent by various policies of the welfare state. Post-Second World War social welfare policy in Canada has generally been closer to that of the United States than of Western Europe. In contrast to many Western European countries, Canada has not followed a full employment policy. Unlike the United States, however, Canada has a universal health care system and somewhat more generous social service programmes such as unemployment insurance. When in power, the two major Canadian political parties (at the federal level) have generally followed a pragmatic and centrist approach to social welfare provision, although the election of a more neo-conservative government in 1984 and its re-election in 1988 raised concerns about continued federal support for various social assistance programmes (Mishra 1990: 70–9). These concerns were eased in 1993 with the return to power of a more moderate Liberal government although the Liberals have also instituted changes in social welfare provision in order to control the federal deficit. More importantly, however, the election in 1995 of an avowedly neo-conservative government in Ontario, the province of which Toronto is the capital and major urban centre, has accelerated concern about a substantial dismantling of the welfare state and the implications of this for the least well-off in the Toronto area.

In the following sections I shall expand on a number of these and related themes. The second section reviews the nature of the Canadian welfare state, especially compared to those of the United States and a number of European countries. Consideration is also given to the implications of recent political events, including the federal government's ongoing review of welfare programmes and the change of government in Ontario. In the third and fourth sections recent changes in Toronto's economic structure and immigration patterns are discussed. These sections provide the context for a detailed consideration of the dynamics of social–spatial polarisation in Toronto, dealt with in the fifth section. Together, these sections provide the background for an elaboration of whether, where and to what extent these factors affect the chance of becoming socially excluded in Greater Toronto.

The Canadian welfare state: maintaining the status quo?

The essentials of the modern welfare state in Canada were outlined immediately following the Second World War in two documents, the White Paper on Employment and Income and the Green Book. The White Paper on Employment and Income emphasised the importance of exports, especially staple goods, as a means of enhancing employment and incomes. To a considerable extent Canada has remained a resource-based and branch-plant economy, thereby enhancing the country's vulnerability to decisions made elsewhere. The Green Book laid the foundation for a number of social assistance programmes including unemployment insurance, old age pensions and national health insurance (Lightman and Irving 1991: 67–8). Most of the latter were implemented in the 1950s and early 1960s so that by the mid-1960s the key elements of the welfare state, as laid out in the Green Book, were in place. Implementation of these programmes was eased financially by increased economic growth and rising tax revenues during the 1960s and early 1970s.

Economic circumstances changed dramatically following the OPEC oil embargo in 1973. The outcome was a serious decline in the tax base that had supported new social programmes and rising government debt. Despite rising deficits, however, there has been general support among Canadians for many of the social programmes that were put in place in the 1960s and governments have generally responded to public opinion. Although no new social programmes have been introduced in the last two decades, there have been no major reductions in social programmes either, at least by the federal government (Mishra 1990: 76–9).

Support for social programmes came to the fore in the 1984 election and 1988 re-election of the Conservative government. The 1988 election was fought on the issue of a proposed free trade agreement with the United States. Although the Conservative government argued that Canada's social policies would not be affected by the free trade agreement, the opposition expressed concern that under free trade many of Canada's more generous social programmes would be eliminated or severely curtailed in order to improve the competitive position of Canadian industry, particularly against firms employing non-union labour in the southern United States (Lightman and Irving 1991: 75). Subsequently, the Conservatives won the election, implemented the free trade agreement, imposed large tax increases to cover increased social welfare expenditures and introduced a repayment provision based on income for two universal social welfare programmes, family allowances and old age security. In contrast to Britain, however, there has been no widespread dismantling of the welfare state and the Conservative government retreated from an assault on universal social programmes when it became clear that these were favoured by a majority of the electorate (Mishra 1990: 73–5). Indeed, O'Connor (1989) has argued that universal social programmes, particularly health insurance, have encouraged public resistance to extensive dismantling of Canada's welfare state.

There are signs, however, that public opinion has shifted in the 1990s with the move of the ruling Liberal Party at the federal level to a more fiscally conservative position, the emergence of a more right-wing Reform Party, rather than the Conservatives, as the major conservative opposition, the election of more neo-conservative provincial governments in Alberta and Ontario, and a relative decline in the fortunes of the New Democratic Party, Canada's social democrats.

In global terms, Canada has been categorised as a liberal welfare state (Esping-Andersen 1989: 25). Social security assistance is usually means-tested, benefits are relatively modest and the low-income working-class recipients tend to be residualised from the rest of society. Much more emphasis has been placed on universal social programmes such as health and education than on full employment and redistributive programmes such as unemployment insurance, old age security and other forms of social assistance. Compared to other OECD countries, Canada spends a relatively large amount on public health care and education but comparatively little on services such as job training, social housing and social transfer expenditures (O'Connor 1989, Coyne 1994).

In comparison to seven other North American and European countries Canada ranked fifth during the latter half of the 1980s in social security benefits expenditures as a percentage of GDP, considerably above the United States, about even with Britain and Belgium and considerably behind Germany, France, the Netherlands and Sweden (Table 5.1).[3] In part, Canada's relatively high expenditure on social security benefits compared to the United States relates to higher unemployment rates in Canada. Canada's total unemployment rate in 1992 was 11.2 per cent compared to 7.3 per cent in the United States (Table 5.1).[4] Indeed, Canada's unemployment rate was higher than that of any of the other seven countries shown in Table 5.1. Obviously, attempts at achieving full employment in Canada have not been very successful. The unemployment rate has dropped recently but still remains between 9 and 10 per cent.

Of more importance, however, in explaining variations between Canada and the United States are the substantial differences in values and cultures, particularly associated with individualism and collectivism, between the two countries. Greater emphasis on collectivism in Canada has led to a higher level of government intervention in the welfare state (e.g. Goldberg and Mercer 1986, O'Connor 1989). This is particularly evident for public health insurance, public education expenditures and unemployment insurance benefits. In the United States, these programmes are generally less universal and not as generous.

Compared to the other countries in Table 5.1 it appears that social spending in Canada has had relatively little impact on income redistribution. The ratio of incomes for the highest 20 per cent of households to the lowest 20 per cent is second only to the United States and considerably above more social democratic countries such as Sweden. Viewed over time, the social transfer system has had little impact on the redistribution of incomes in Canada. Table 5.2 indicates that there was remarkable stability in the distribution of

Table 5.1 Social security benefits expenditures, unemployment rates, income inequality and debt interest (1979 and 1990) for selected countries

Countries	*Social security benefits/ % of GDP 1985–90 (UN)*	*Benefit expenditure/ % of GDP 1991 (ILO)*	*Social security and other transfers/ % of GDP 1990 (OECD)*	*Unemployment rate, 1992*	*Income inequality, 1980–91*	*Debt interest, 1979/ % of GDP*	*Debt interest, 1990/ % of GDP*
United States	12.6	12.0	11.5	7.3	8.9	2.8	5.2
United Kingdom	17.0	18.3	13.7	9.9	6.8	4.4	3.4
Canada	18.8	18.6	12.8	11.2	7.1	5.0	9.4
Belgium	19.8	na	24.8	7.8	4.6	5.3	10.7
Germany	23.0	23.6	19.3	4.8	5.7	1.7	2.6
France	26.1	na	23.5	10.2	6.5	1.4	3.1
Netherlands	28.7	28.4	29.1	6.8	5.6	4.2	6.7
Sweden	33.7	35.2	21.2	4.8	4.6	4.1	5.8

Source: United Nations Development Programme (1994) *Human Development Report 1994*, New York: Oxford University Press

Note
Income inequality refers to the ratio of incomes for the highest 20 per cent of households to the lowest 20 per cent.

Table 5.2 Percentage distribution of income for families by quintiles, 1971, 1981 and 1992, after transfers and before tax (total money income), Canada

	Lowest	*Second*	*Middle*	*Fourth*	*Top*
1971	5.6	12.6	18.0	23.7	40.0
1981	6.4	12.9	18.3	24.1	38.4
1992	6.3	12.2	17.8	24.0	39.7

Source: Statistics Canada, Household Surveys Division

income by quintiles (after transfer payments) between 1971 and 1992. In one way, this is not surprising because of the relative stability of social programmes during these two decades. As Banting (1987: 331) notes, however, this stability in income distributions may have been due to the redistributive function of the state, especially during periods of economic stagnation. Consequently, through income redistribution programmes, the state may have played an important role in moderating the negative effects of economic downturns.

Recently, the Canadian government has announced an extensive review of social programmes. The objective is to improve job training, target programmes more directly to the needy and make social programmes more efficient. A major impetus for this review is Canada's growing debt. As noted in Table 5.1, Canada's interest payment on the national debt increased dramatically between 1979 and 1990, and is now second to Belgium among the countries shown in the table. In addition, Canada's tax burden, as a percentage of GDP, has risen considerably above that of the United States. At the heart of the debate are two issues. The first is the universality of social programmes versus targeting programmes to the most impoverished in society. McQuaig (1993: 28) has suggested that this is the basic difference between the European and American models of social welfare. The most vociferous critics of proposed changes to the Canadian system argue that a shift to more targeted programmes has the potential of eroding political support for these programmes and further stigmatising those who receive social payments (e.g. McQuaig 1993). On the other side are those who argue that the current welfare state precipitates high government spending that encourages dependency and erodes productivity and economic growth (e.g. Watson 1994). Still others (e.g. Mishra 1990) argue that even neo-conservative governments have found it difficult to dismantle existing universal programmes because of the popular support they enjoy. Therefore, the potential for dismantling universality in Canada may not be as great as some critics suggest.

The second major issue is the willingness of citizens to pay higher taxes to sustain the welfare state at current levels. Potentially, this is a more serious threat to welfare provision and the well-being of the poorest in society. Richards (1994) has suggested that marginal tax rates above 50 per cent tend to attract a much higher degree of tax evasion. He also points out that voters are likely to tolerate public expenditures of 40 per cent of GDP but once the figure goes much above 50 per cent governments tend to be defeated at the next electoral opportunity. Public sector debt in Canada is now above the upper level of tolerance and although the current Liberal government has a strong mandate, it is not likely to annoy the electorate with substantial tax increases in order to reduce the debt. The outcome will be further reductions and more specific targeting of social spending. In contrast to previous years, when taxes were increased to cover increased costs of social transfer benefits, the finance minister has argued recently that the objective must be to reduce government expenditures and hold personal income tax rates at current levels (Department of Finance, Canada 1995).

The federal government is the major shaper of the welfare state in Canada but it is not the only one. Although the federal government establishes broad policy and makes financial transfers to the provinces, the provinces and municipalities also collect taxes for social welfare spending and have considerable control over the way in which a broad range of social services and social assistance are allocated (Laws 1994). For example, social assistance is under provincial jurisdiction and rates vary across the country (Mishra 1990: 75). In Ontario the more restrictive policies of the federal government on social spending have been exacerbated dramatically by the policies of a neo-conservative government. Soon after assuming office in 1995 the new government announced a number of deficit-reduction measures that have had a direct effect on the least well-off in the province. These include a 21.6 per cent reduction in payments received by all welfare recipients except the elderly and disabled, cancellation of employment training programmes and job creation initiatives and suspension of all new social housing construction. Although social assistance payments have always been relatively low in Canada, this is clearly a retreat from the status quo and the centrist position that various governments in Ontario have adhered to since the Second World War. Low-income people in the Toronto area have been most severely affected because of the relatively high cost of living in Toronto.

At the local political level there tends to be an uneven delivery of services by a complicated mix of public and private agencies operating at various jurisdictional levels. In Toronto, the creation in 1953 of Metropolitan Toronto, a federation of local municipalities, tended to even out the provision of major social services across this part of the city although a number of facilities such as hospitals, day care centres and emergency shelters tend to be concentrated in the central part of the City of Toronto (Figure 5.1).[5] Since 1953, however, the built-up area of the Toronto region has grown considerably beyond the borders of Metropolitan Toronto and although regional governments have been formed in these rapidly growing areas they and their constituent municipalities are generally less receptive to the establishment of 'visible' social services than the City of Toronto and some of the other municipalities in Metropolitan Toronto.

Recent changes in economic structure

During the past few decades Toronto has experienced many of the changes that are associated with post-industrial restructuring: a decline in manufacturing and an increase in service activities, a rapid increase in producer services and decentralisation of manufacturing and routine office functions to the suburbs (Murdie 1996). Between 1961 and 1992 the share of Toronto's labour force in manufacturing declined continuously from 29.7 per cent in 1961 to 18.1 per cent in 1992 (Table 5.3). During the same period employment in community, business and personal services (CBP services) increased from 21.2 per cent to 36.5 per cent and employment in finance, insurance and real estate went up from 6.6 per cent to 10.0 per cent. These trends are similar to employment shifts in other post-industrial metropolises (e.g. Sassen 1991, 1994a, Buck, Drennan and Newton 1992, Fainstein and Harloe 1992, Gordon and Sassen 1992). One of the key elements in the development of the post-industrial metropolis is the expansion of producer services and financial activities (Sassen 1994a: 55–65). As noted previously, Toronto has become the pre-eminent financial centre in Canada and is now the dominant 'command and control' metropolis in the country. Almost 200 of Canada's 500 largest companies had their head offices in the Toronto area in 1992 (Financial Post 1993: 100–18).

Although much of the decline in Toronto's manufacturing activity has resulted from technological change, Toronto is also vulnerable to decisions made outside the country. According to Semple (1988: 347), 72 per cent of revenues from manufacturing in Toronto in 1985 came from foreign-owned firms. Multinational firms can easily shift capital and production from one

Table 5.3 Toronto Census Metropolitan Area: percentage employment by industry, 1961–92

Industry	*1961*[a]	*1971*[a]	*1981*[a]	*1986*[a]	*1990*[b]	*1992*[b]
Manufacturing	29.7	25.3	23.8	22.7	19.8	18.1
Construction	6.5	6.1	5.4	5.5	6.2	5.0
TCU	8.7	7.5	8.0	7.4	7.5	7.5
Trade	18.6	16.7	18.1	18.2	17.6	17.0
FIRE	6.6	6.8	8.5	8.5	9.8	10.0
CBP services	21.2	24.1	29.6	31.8	34.1	36.5
Public admin.	5.4	5.4	5.1	4.8	5.0	5.1
	96.7	91.9[c]	98.5	98.9	100.0	99.2

Notes
a Census of Canada. Base population is the total labour force.
b Statistics Canada, Labour Force Survey. Base population is the employed labour force.
TCU = Transportation, communications and utilities.
FIRE = Finance, insurance and real estate.
CBP = Community, business and personal services.
c Excludes 7.6 per cent unspecified or undefined.
Primary industries and unclassified industries are not shown.

location to another, a phenomenon that has been demonstrated for Toronto by Norcliffe, Goldrick and Muszynski (1986). Enactment of free trade agreements with the United States and subsequently Mexico in the late 1980s and early 1990s has increased the risk that more manufacturing activity will leave Toronto.

Like many other post-industrial cities, Toronto has experienced a decline in manufacturing activity in the central city and an increase in the suburbs, particularly the regional municipalities of Durham, York, Peel and Halton.[6] In the City of Toronto, the number of manufacturing firms declined from just over 2,000 in 1971 to 1,700 in 1986 and 1,500 in 1993. Employment in manufacturing and warehousing declined even more dramatically, to about 24,000 employees in 1993 compared to 65,800 in 1986 and 82,000 in 1971 (Metropolitan Toronto Planning Department 1992: Figure 2.5, 1995: Figures 2.9 and 2.12). In contrast, the number of firms in the four outer regions increased from 1,500 in 1971 to 3,300 in 1986 and employment in manufacturing and warehousing during the same period increased from 98,900 to 141,500 (Metropolitan Toronto Planning Department 1992: Figure 2.5).

In contrast to manufacturing activities, office space has increased dramatically in the central area of Toronto. There was 2.5 times more office space in the Central District of the City of Toronto in 1991 than in 1971. Many of the firms that have located here are head offices of banks, trust companies, investment firms and insurance companies, firms that often require face-to-face communication (Gad 1985: 338). As Sassen (1994a: 85) notes in the context of Toronto's financial district, 'the risk, complexity, and speculative character of much of this activity raises the importance of face-to-face interaction'. This is interaction that cannot be totally replaced by modern telecommunications (Coffey 1994: 76–7). In contrast, routine 'back office' functions such as data processing and credit card and accounting operations, as well as the head offices of foreign-controlled companies, have tended to locate in suburban office parks in close proximity to good highway access. However, while the number of offices in the outer suburbs has increased dramatically in the last two decades, the central area still accounts for over half of Toronto's office floor space (Metropolitan Toronto Planning Department 1992: Figure 4.2).

Like many developed countries, Canada entered a severe recession in the 1990s that has impacted dramatically on the labour force and unemployment rates. The outcome is evident in job statistics comparing the 1984 to 1990 period with 1990 to 1992. Between 1984 and 1990 Toronto's labour force increased by about 300,000 while in the succeeding two years the labour force declined by 156,000 (Canada Employment and Immigration Centre 1992). Loss of jobs in the manufacturing sector between 1990 and 1992 accounted for almost 40 per cent of the total or 60,000 jobs. In contrast, jobs in the producer services sector were much less severely affected.

Unemployment rates increased dramatically during this period, from 5.1 per cent in 1990 to 11.5 per cent in 1992 (Canada Employment and Immigration

Centre 1992: Table 1). Youths were particularly affected, with unemployment reaching 19.1 per cent in 1992 among the 15–24-year age group. As well, the percentage of total employment that is part-time increased from 10.1 per cent in 1983 to 19.3 per cent in 1992 (Metropolitan Toronto Planning Department 1993b). Although some part-time employment is voluntary, the Committee of Planning and Coordinating Organizations (1992: 40) notes that there has been a considerable increase in involuntary part-time work in Toronto in the past few years. Such jobs are often characterised by low wages and few benefits and are particularly held by women and young people under 25.

Recent changes in immigration

During the last three decades there has also been a dramatic shift in the origins of Toronto's immigrants. Until the late 1960s most of Toronto's immigrants were from Britain and other European countries. This was due to Canada's policy of giving preference to 'White' immigrants from other Commonwealth countries, continental Europe and the United States. In 1967 the 'preferred' system was replaced by a 'point system' which treated immigrants in the same way, regardless of origin. The impact on the origin of immigrants coming into the country by period of immigration has been dramatic. As Hiebert (1994: 255) notes, 'the flow of immigrants to Canada has been profoundly internationalised'.

The internationalisation of Toronto's immigrant population is shown in Figure 5.2. Of particular note is the dramatic decline in immigration from Europe during the post-Second World War period, but especially following the shift in immigration policy in 1967. Of immigrants living in Toronto in 1991, almost 64 per cent of those who arrived in the 1966–70 period originated from Europe while only 36 per cent of those who arrived in 1971–5 were from Europe. This figure also indicates the increased heterogeneity of Toronto's immigrant population, especially since the early 1970s. Of particular note is the substantial increase in immigrants from South and East Asia (primarily Hong Kong, India and Sri Lanka). Almost 40 per cent of Toronto's 1991 immigrant population who arrived in the previous five years came from South and East Asia. Also shown in Figure 5.2 is the relatively large in-migration of Caribbeans during the early 1970s and the recent inflow of immigrants from South East Asia (e.g. Vietnam and the Philippines), Africa and Central and South America.

The immigrant flow into Toronto also represents a wide spectrum of economic groups, including refugees who are admitted on humanitarian grounds, those joining families already in Canada, business people with entrepreneurial skills and capital to invest and independents admitted through the points system or persons with relatives in Canada who agree to provide financial support if needed. Points are awarded particularly for educational achievement, labour market skills and language proficiency. Of the 72,000 immigrants and refugees destined for Toronto in 1993, approximately 41 per cent were in the family class, 32 per cent

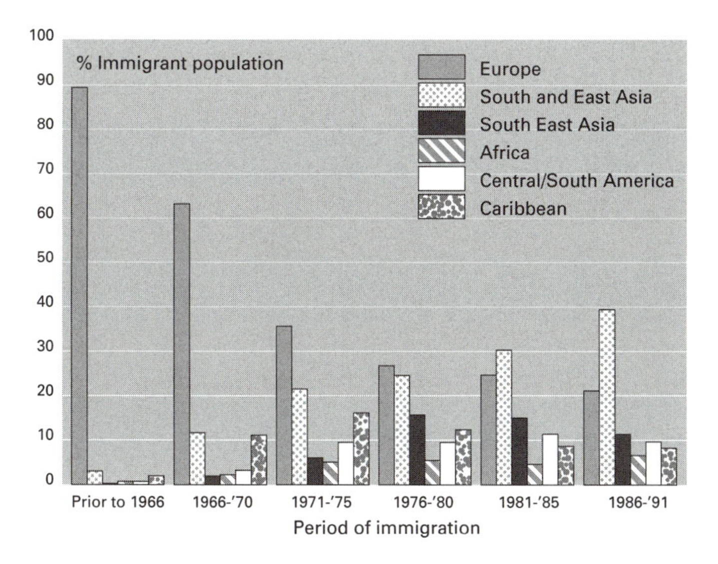

Figure 5.2 Place of birth by period of immigration, Toronto Census Metropolitan Area, 1991

Source: Census of Canada, 1991

were independents, 15 per cent refugees and 9 per cent in the business category (Citizenship and Immigration Canada 1994a: 12). The federal government's intention in the future is to attract more business and independent class immigrants, rather than family class, because of the likelihood that these groups will make a greater contribution to the economy and require less welfare assistance (Citizenship and Immigration Canada 1994b: 5–16).

The dynamics of social–spatial polarisation in Toronto

Shifts in occupational structure: polarisation or professionalisation?

As noted by recent critics, social polarisation is not an easy idea to conceptualise or operationalise (Hamnett 1994a, Kempen 1994). Criticism has centred on the ambiguity of the term, as originally proposed by Sassen (1991), and the view that an increasingly bimodal division of the labour force into low-skilled and high-skilled jobs may be limited to a few cities, especially in the United States, and particularly cities that have experienced a large inflow of immigrants from low-wage countries. Indeed, Hamnett (1994a) has suggested that professionalisation, the shift of large numbers of the labour force into managerial and professional jobs, may be the dominant process in most large post-industrial cities.

In relative terms, Toronto's occupational structure changed substantially during the 1976 to 1992 period (Table 5.4). In 1976, 25.1 per cent of the labour force was employed in managerial and professional jobs while by 1992 this figure increased to 37.1 per cent (Census of Canada and Statistics Canada, Labour Force Survey). In contrast, the proportion of the labour force in most other categories declined. Even service jobs declined slightly, from 12.1 per cent of the total labour force in 1976 to 11.9 per cent in 1992. Not unexpectedly, manufacturing occupations declined most dramatically, from 15.8 per cent of the labour force in 1976 to 11.2 per cent in 1992.

In absolute terms, Toronto's labour force increased about 300,000 between 1984 and 1990, the period immediately before the recession of the 1990s (Canada Employment and Immigration Centre 1992). Of these new jobs, almost 50 per cent were managerial and professional. A further 20 per cent were clerical and about 17 per cent were in sales. Only 3 per cent were service occupations and, not unexpectedly, only 1.5 per cent were manufacturing. Although caution is needed because many clerical and sales jobs are low-waged, these figures generally support Hamnett's professionalisation hypothesis. In this sense, Toronto corresponds to the general pattern whereby professionalisation has been the primary source of occupational change in cities where a substantial producer services sector has developed.

Changing income distributions: a shift to the executive inner city?

Income inequality, particularly by quartiles, quintiles or deciles, is another way of measuring the degree of social polarisation within a metropolitan area. Unfortunately, income figures in this form for Toronto are very difficult to obtain. The only detailed analysis for the entire Toronto area is based on 1970 and 1980 household income data (City of Toronto Planning and Development Department 1984). Additional evidence for the central core area is available

Table 5.4 Toronto Census Metropolitan Area: percentage employment by occupation, 1976 and 1992

Occupation	*1976*[a]	*1992*[b]
Managerial and professional	25.1	37.1
Clerical	22.9	18.5
Sales	12.1	10.2
Service	12.1	11.9
Processing/machining/product fabrication	15.8	11.2
Construction	5.7	3.9
Transportation equipment operation	3.4	2.8
Material handling	2.7	3.8

Notes

a Census of Canada. Base population is the total labour force.

b Statistics Canada, Labour Force Survey. Base population is the employed labour force.

to 1985 (Ram, Norris and Skof 1989: 33). Therefore, it is not possible to say much about income distribution or redistribution among Toronto households beyond 1980. It is possible, however, to say something about changes in the geography of income distributions in Toronto. Indeed, there has been a considerable debate, especially in Toronto, about the impact of inner-city gentrification on social polarisation and the distinction between gentrification and the more general concept of urban revitalisation (e.g. Ley 1992, Bourne 1993a, 1994).

Although individual income figures are not readily available, it is possible to comment on city–suburb contrasts and spatial polarisation between small areas within the city. In contrast to many United States cities, there is considerable reason to believe that the city–suburb income gap in Toronto is narrowing and that areas within Toronto, especially the central city, have become more spatially polarised over time. To a large extent, this assertion relates to changes in central city housing stock. These changes include the loss of affordable rental housing and the production of new housing, much of which is targeted to higher-income groups.

The loss of affordable rental housing in the inner city has come from two major sources: (1) the demolition of older private rental buildings and, in some cases, their conversion to other uses including owner-occupied condominiums,[7] and (2) the deconversion of houses with flats to single family units. Traditionally, it has been assumed that deconversions have resulted largely from the pressures of gentrification. Murdie and Northrup (1989), however, found in an interview study in the central city that the largest percentage of deconverted properties were owned by Portuguese, many with large families who needed more space and no longer had a mortgage. Nevertheless, it should be stressed that several areas of the central city have experienced gentrification or middle-class resettlement, although the exact number of units affected is unknown (e.g. Caulfield 1994).

The production of new housing in Toronto has been primarily socially assisted housing and condominium units. In many instances, these new residential developments emerged out of the central city's de-industrialised landscape or 'greyfield' sites of old industrial, railway and port lands (Bourne 1992: 78–9, Caulfield 1994: 77). New social housing has taken the form of non-profit and co-operative housing. In contrast to traditional forms of public housing, non-profit and co-operative housing allows for a greater mix of incomes and thereby attempts to minimise social polarisation (Dreier and Hulchanski 1993). Condominium home ownership has only been permitted in Ontario since 1967 but it has become an increasingly popular form of tenure, especially in the city core and the surrounding area of mixed use and high-density development. Most condominiums in the central core are luxury units, the prices of which escalated dramatically in the latter part of the 1980s (Preston, Murdie and Northrup 1993: 289). A questionnaire survey of resident and non-resident condominium owners in the City of Toronto revealed that resident condominium owners had a much higher average family income than the average household in the City of Toronto (Preston, Murdie and Northrup 1993: 287).

Before the results of the income study are discussed, it is important to define the geographic scope of this and subsequent analyses. Reference will be made to four areas: the inner city, the central city, the inner suburbs and the outer suburbs (Figure 5.3). These have been defined largely according to period of development. The inner city is defined by the boundary of Toronto's central area plan. Spatially, this area includes the core and surrounding area of mixed land use and residential development. The central city, which also includes the inner city, is defined as the City of Toronto. The City of Toronto was incorporated in 1834, expanded rapidly in the first thirty years of the twentieth century and then levelled off at slightly less than 700,000 population. The population of the City of Toronto has fluctuated since then but unlike many United States central cities there has been no major post-Second World War abandonment or decline in population (Bourne 1992).

The inner suburbs comprise the five constituent municipalities of Metropolitan Toronto outside the City of Toronto that were largely built up between the end of the Second World War and the early 1970s. Between 1986 and 1991 population in this area increased by a modest 3.8 per cent. The outer suburbs include the rapidly growing suburban municipalities beyond Metropolitan Toronto but within the Toronto Census Metropolitan Area (CMA). Population in this largely middle-class, family-oriented area increased by 27 per cent between 1986 and 1991. As the population of the CMA

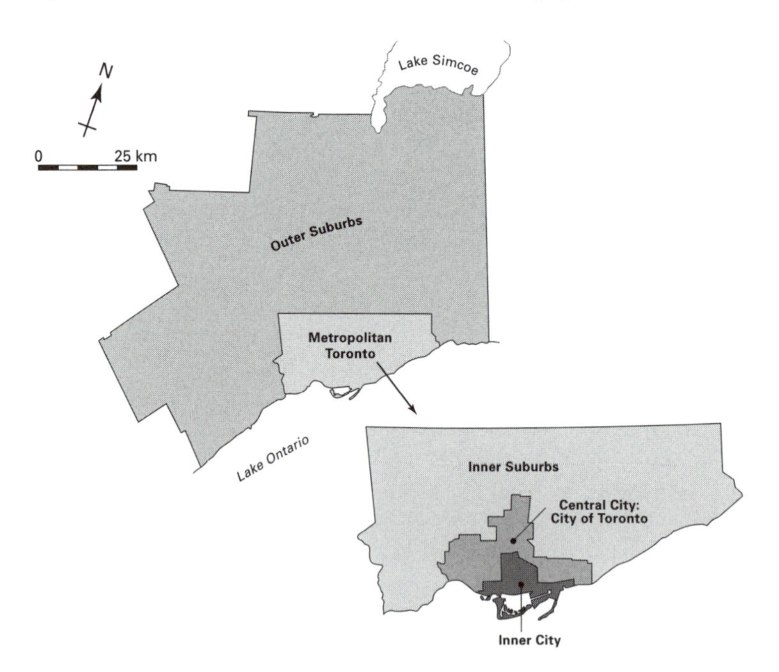

Figure 5.3 Toronto Census Metropolitan Area: inner city, central city, inner suburbs, outer suburbs

expanded, Metropolitan Toronto's share of the CMA population dropped from 80 per cent in 1971 to 58 per cent in 1991. Metropolitan Toronto is now the core of a much larger metropolitan area.

Income ratios for the inner city and the central city versus the CMA were calculated for 1970 and 1990.[8] The two time periods correspond with census years: 1970 represents the beginning of large-scale middle-class upgrading in the central city and 1990 is the most recent year for which data are available.[9] The results show that there is still an income gap between the Toronto CMA and both the inner city and the central city, at least for average household income. For the inner city, however, the gap narrowed from 71 per cent of CMA income in 1970 to 77 per cent in 1990. For a more narrowly defined area containing the central business district and its immediate vicinity, Bourne (1994: 567) found a more substantial increase in median household income. Incomes for this area as a ratio of the CMA increased from 0.81 in 1970 to 1.03 in 1990. Clearly, changes in the central city housing market, particularly the construction of new luxury condominiums, have had some impact in modifying the gap. For the central city, average household income was about 90 per cent of the CMA average for both 1970 and 1990. This differential is considerably less than similar figures for nearby United States cities such as Detroit (60 per cent) and Cleveland (61 per cent) (Rusk 1993: 33). Bourne (1993b) argues that we should not be surprised at the lower income levels in the central city for at least two reasons. One concerns the presence of persons in the central city such as immigrants, students and recent university graduates whose current income is low. The other is the continued availability of low-cost housing, including various forms of socially assisted housing, in the central city.

Income inequality or polarisation can be identified by a variety of measures. The measure used here is the interquartile (and interdecile) range for household income, 1970 and 1990, using data for census enumeration areas in the inner city.[10] The increased discrepancy between rich and poor areas within the inner city is shown by a comparison of the quartiles in Table 5.5. In 1990 dollars, the interquartile range increased from $14,425 in 1970 to $24,951 in 1990. When examined by deciles, the polarisation is even more extreme. The range between the bottom and top deciles in 1990 dollars increased from $31,023 in 1970 to $59,900 in 1990. Put another way, the ratio between the top and bottom deciles increased from 2.81 in 1970 to 4.07 in 1990. This suggests a scenario where household incomes for areas in the lowest decile and quartile barely kept up with inflation while those in the upper quartile and decile vastly exceeded inflation. Not unexpectedly, areas in the lowest decile are primarily public housing developments while those in the highest decile are new luxury condominium developments. The evidence suggests considerable reinvestment, income redistribution and spatial polarisation between rich and poor areas in this part of the city.[11] It also suggests that Toronto may be on its way towards becoming an executive inner city but with persistent pockets of poverty throughout.

Table 5.5 Interquartile and interdecile ranges, Toronto inner city, 1970 and 1990 (income in 1990, $)

	1970	*1990*
Quartiles		
Lower	25,994	31,892
Upper	40,369	56,041
Interquartile range	14,425	24,951
Ratio: upper/lower	1.56	1.76
Deciles		
Lower	17,156	19,495
Upper	48,179	79,395
Interdecile range	31,023	59,900
Ratio: upper/lower	2.81	4.07

Source: Census of Canada, 1971 and 1991. Calculations by the author based on enumeration area data

The declining inner suburbs?

With large-scale middle-class upgrading in the inner city, and continued middle-class suburbanisation of the outer suburbs, there is some evidence that the inner suburbs have declined in socio-economic terms. The differences between the central city, inner suburbs and outer suburbs, and changes over time, are shown by indices of over- and under-representation in Table 5.6. These indices are calculated in the same way as those for income in the second section. The central city, inner suburbs and outer suburbs are defined in the same way as before. Figures are provided for the incidence of low-income families, low educational achievement, lower levels of occupational status and unemployment. Occupational status and unemployment are of particular interest because of the availability of data for a twenty-year time horizon (1971, 1981, 1991) compared to a decade (1981, 1991) for low-income families and educational achievement.[12]

In relative terms, there is a clear gradient in the ratio of low-income families from the central city to the outer suburbs. The central city percentage was about 1.5 times the CMA figure in both 1981 and 1991 while the outer suburbs had about 60 per cent of the CMA percentage of low-income households in both years. These figures confirm the continued pockets of poverty in the central city, in spite of an increased shift towards an executive city. Of particular interest, however, is the relative increase in the percentage of low-income families in the inner suburbs, from 1.04 times the CMA figure in 1981 to 1.26 in 1991.[13] The educational achievement figures also show a gradient from the central city to the outer suburbs and, as with low-income households, the ratio for low educational achievement increased over the decade in the inner suburbs.

Compared to the central city and the outer suburbs, the inner suburbs had the highest ratio for manufacturing workers in 1991. Also, while the ratio for manufacturing employees declined in both the central city and the outer suburbs

Table 5.6 Toronto Census Metropolitan Area: ratios of over- and under-representation, income, education and occupation, for the central city, inner suburbs and outer suburbs (1971, 1981 and 1991)

Variables	*Central city ratio*	*Inner suburbs ratio*	*Outer suburbs ratio*
Income			
(% low-income families)			
1981	1.51	1.04	0.62
1991	1.54	1.26	0.59
Education			
(population 15 years and over not attending school full-time)			
% less than grade 9			
1981	1.32	1.03	0.69
1991	1.29	1.12	0.74
Occupation (labour force)			
% manufacturing			
1971	0.93	1.00	1.12
1981	0.84	1.04	1.04
1991	0.82	1.12	0.95
% service			
1971	1.32	0.89	0.82
1981	1.32	0.97	0.82
1991	1.37	1.04	0.84
% unemployed			
1971	1.26	0.93	0.78
1981	1.19	1.01	0.85
1991	1.28	1.15	0.78

Source: Census of Canada, 1971, 1981 and 1991

between 1971 and 1991, it increased in the inner suburbs during this time period from 1.00 to 1.12. Similarly, the index for service employees, although not as high as the central city, increased from 0.89 to 1.04. Overall, the inner suburbs appeared to be declining in status between 1971 and 1991. This view is further confirmed by the ratios for unemployment. Like the ratios for low-income households, these ratios declined from the central city to the outer suburbs, although differences between the three areas are not as extreme. What is noticeable, however, is the increase in the ratio in the inner suburbs from 0.93 in 1971 to 1.15 in 1991 while the other two areas had virtually the same ratios in both years.

This relative decline in status of the inner suburbs results primarily from two phenomena. One concerns the changing geography of ethnicity in Toronto, especially the increased number of new immigrants and refugees who have found accommodation in lower-rent areas of the inner suburbs. This is perhaps the major factor. The other is the number of low-income families living in public housing developments, many of which were built on greenfield sites in the inner suburbs (especially North York and Scarborough) during the 1960s and 1970s. This housing was often developed in close proximity to low-cost private rental accommodation, thereby increasing the number of relatively low-rent units in these areas and the concentration of low-income households.

The changing geography of ethnicity

As noted earlier, the source countries of Toronto's immigrant population have changed dramatically during the past three decades. So, too, has the spatial differentiation of ethnic groups within the Toronto CMA. Table 5.7 presents some details of these trends for 1971 to 1991, using the index of over- or under-representation for eight ethnic groups plus recent immigrants. The ethnic groups selected for analysis represent the host British population and the Jews, Germans, Italians, Portuguese, Blacks, South Asians and Chinese.[14] Like the British, the Jews are well established and have achieved a high level of upward social mobility. When compared to the British, however, they have maintained a high level of spatial segregation. The Germans arrived shortly after the Second World War, are generally engaged in skilled jobs and exhibit the smallest degree of segregation. The Italians came primarily during the 1950s and 1960s, generally with low levels of education and occupational skills. The Portuguese followed the Italians in the 1960s and 1970s and have not yet suburbanised to the same extent. In contrast to the European groups, the Blacks (initially from the Caribbean and more recently from various African countries), South Asians and Chinese arrived in Toronto primarily following liberalisation of immigration laws in the 1960s and 1970s. The recent immigrants are those who arrived in the period two and a half years before the census.

Indices of dissimilarity indicate considerable variation in the degree of residential segregation between these groups and the British population. The Jews have the highest index value (79 in 1991) and the Germans the lowest (19 in 1991) (Balakrishnan and Hou 1995: 24). For both groups, the indices have changed very little since 1971 (Kalbach 1990: 98, Davies and Murdie 1993). The Portuguese were quite highly segregated in both 1971 (68) and 1981 (62). Although a comparable figure is not available for 1991, it is unlikely that the level of segregation of this group has dropped dramatically since 1981. The Italians also exhibit a relatively high level of segregation that has not dropped much over time (57 in 1971 and 56 in 1991). The three most recently arrived groups all exhibit indices in the 50 to 60 range. Of these groups, the South Asians have the highest index (66 in 1991), followed by the Chinese (57) and the Blacks (52). Indices for the Chinese and Blacks have increased slightly since 1971 and the indices for all of these groups increased marginally between 1986 and 1991.

One of the major contrasts between Toronto and most United States metropolitan areas is the difference in degree of racial segregation, especially for the Black population. Blacks in many United States cities tend to experience extreme

Table 5.7 Toronto Census Metropolitan Area: ratios of over- and under-representation, ethnic groups and recent immigrants for the central city, inner suburbs and outer suburbs (1971, 1981 and 1991)

Ethnic origin (% population)	*Year*	*Central city ratio*	*Inner suburbs ratio*	*Outer suburbs ratio*
British	1971	0.80	1.00	1.30
	1981	0.84	0.95	1.19
	1991	0.88	0.90	1.14
Jewish	1971	0.83	1.43	0.13
	1981	1.01	1.36	0.35
	1991	1.08	1.24	0.73
German	1971	0.75	1.00	1.32
	1981	0.78	0.90	1.32
	1991	0.89	0.91	1.13
Italian	1971	1.21	1.12	0.40
	1981	0.78	1.22	0.76
	1991	0.56	1.13	1.05
Portuguese	1981	3.00	0.50	0.51
	1991	2.58	0.55	0.84
Black	1971	1.54	0.99	0.31
	1981	0.74	1.26	0.69
	1991	0.66	1.43	0.71
South Asian	1981	0.62	1.22	0.86
	1991	0.39	1.14	1.11
Chinese	1971	2.49	0.53	0.24
	1981	1.82	0.99	0.46
	1991	1.39	1.21	0.64
Recent immigrants	1971	1.84	0.82	0.36
	1981	1.49	0.98	0.70
	1991	1.05	1.31	0.68

Source: Census of Canada, 1971, 1981 and 1991

Note
Ethnic origin refers to the ethnic or cultural group(s) to which the respondent's ancestors belong. Recent immigrants are those who arrived in the period two and a half years before each census.

segregation with indices in the 70 to 80 range (e.g. Massey and Denton 1993: 64) compared to about 50 in Toronto. Balakrishnan and Hou (1995) and Fong (1994) speculate that this difference may result from the lower proportion of Blacks in Toronto, making them a less visible group than the much more numerous Black population in American cities, and the fact that a relatively large proportion of Blacks in Toronto are recent immigrants. These recent immigrants, particularly Caribbean Blacks, are also highly differentiated by skin colour and class (Henry 1994).

As expected, the British host population is least represented in the central city and most represented in the outer suburbs in all three census years. Interestingly, however, the tendency has been towards a relative increase of the British in the central city and a decline in the inner and outer suburbs.

The Jews and Germans have followed similar trends although the Jews were highly represented in the inner suburbs in all three years and have increased their presence in the outer suburbs. All of these groups are of comparatively high occupational status and their relative increase in the more professionalised central city is not surprising.

In contrast to these groups, the Italians have progressively abandoned the immigrant reception area west of the city centre and moved into newer housing in the inner and outer suburbs. Their increased representation in the outer suburbs, from a ratio of 0.40 in 1971 to 1.05 in 1991, has been particularly dramatic. Italians have been especially attracted to home ownership and in 1986 had an ownership rate in Toronto of slightly more than 90 per cent (Ray 1994: 264). Ray (1994: 264) attributes this particularly to the cultural and economic importance that Italians place on the ownership of land. In contrast, the more recently arrived Portuguese are still largely concentrated in the downtown immigrant reception area although they have started to move to the western suburbs of Mississauga and Brampton (Figure 5.1; Teixeira 1995). This is evidenced by an increase in the value of the ratio for the outer suburbs from 0.51 in 1981 to 0.84 in 1991.

The remaining groups, particularly Blacks, are heavily over-represented in the inner suburbs. In 1971 the Black population was strongly concentrated in the central city but by 1981 had become over-represented in the inner suburbs. By 1991, the Blacks had the highest ratio of over-representation in the inner suburbs of any of the groups considered here. In contrast to the Italians, the Black home ownership rate in 1986 was under 40 per cent (Ray 1994: 264). For Black households earning less than $20,000 in 1986 the home ownership rate was 11 per cent compared to 35 per cent for all other ethnic groups (Murdie 1994b: 448). In contrast to other ethnic groups, low-income Black households tend to be a relatively young group dominated by single-parent families. Many occupy public housing, much of which was built on greenfield sites in the inner suburbs during the late 1960s and early 1970s (Murdie 1994a and 1994b).

In contrast to Blacks, South Asians are concentrated in both the inner and outer suburbs and increased their representation in the outer suburbs from 1981 to 1991. Chinese, on the other hand, are heavily over-represented in the central city but the ratio dropped dramatically between 1971 and 1991 as the Chinese increased their presence in the inner and outer suburbs. Finally, the settlement pattern of the recent immigrant group, many of whom are visible minorities without substantial financial resources, also changed dramatically in the 1971–91 period. In 1971 this group was strongly over-represented in the central city but by 1991 recent immigrants became most heavily concentrated in the inner suburbs.

This shift in spatial distribution of more recent immigrant groups results from several factors related to the changing nature of the immigrant population. Business immigrants, for example, have the resources to move directly into high-income areas in the suburbs. Recent immigrants from Hong Kong, in particular, have located initially in the existing Asian communities that have developed in parts of Scarborough, Markham, Richmond Hill and Mississauga (Figure 5.1). Refugees, in contrast, have little choice in housing and many are forced to live in public housing or in low-rent and relatively low-quality private rental and condominium buildings. Many of these are located on inexpensive land in the inner suburbs.

Implications for becoming socially excluded

In the previous sections we have outlined the nature of Canada's social welfare system and at a more local level pointed to a number of structural and spatial shifts that have taken place in the last two decades in Toronto's socio-economic landscape. These include validation of the professionalisation thesis, increased income polarisation in the inner city, the relative decline in socio-economic status in the older inner suburbs, and shifts in the spatial distribution of both the longer-established European immigrant groups and more recently arrived immigrants and refugees from a variety of non-European countries. This section considers the question of whether, where and to what extent these factors affect the chance of becoming socially excluded. Two issues are of particular interest. One concerns the social mobility of immigrants and refugees and the relationship between residential segregation and social exclusion. The other relates to the social residualisation of Toronto's public housing stock. Although treated separately here, these issues are related.

Since the Second World War, Toronto has been the host city for a variety of immigrant groups. During that time there has been a considerable shift in the spatial distribution of these groups, especially from the inner city to the inner and outer suburbs. In general, the immigrant population has also achieved a considerable degree of social mobility. Recent evidence for immigrant households living in Toronto in 1991 indicates that the median income of recently arrived immigrant households (1986–91) was only about two-thirds that of the overall figure for Toronto. However, income increases according to the length of time that households have been in the country (Lapointe and Murdie 1995: Table A6).[15] After fifteen to twenty years in Canada, the median household income for immigrant households begins to equal that of non-immigrant households. Beyond that, except for older immigrant households living on their own, the median income for Toronto's immigrant households exceeds that for non-immigrant households.

There are, of course, differences in social mobility between immigrant groups and within individual groups. For example, recent immigrants from the Caribbean, living in Toronto in 1991, earned only about 88 per cent of the median household income for all immigrant households arriving in the 1986–91 period (Lapointe and Murdie 1995: Table A6). On average, Caribbean households took twenty to twenty-five years to equal the median income of non-immigrant households. In part, this is because many Caribbean households are single-parent families.[16] Henry (1994) also points to the social class divisions in Toronto's Caribbean community and the emergence of a youth underclass frustrated by racial discrimination in education and employment. Many of these youths live with a single parent who may have difficulty in accessing affordable housing in Toronto and may have relatively little left for other essentials after paying monthly housing costs. For this group, Canada's (and increasingly Ontario's) relatively meagre social assistance benefits, combined with the interaction of Caribbean cultural traits and racism by the host society, may have led to what Henry (1994) calls the 'differential incorporation of Caribbean people in Canadian society'.

As noted in the fourth section, there are considerable differences in the spatial segregation of immigrant groups in Toronto. Residential segregation may be a social problem if it also results in social exclusion, discriminatory practices and the lack of access to decent housing or other opportunities. Segregation is not always negative, however. In many instances it provides an important social-psychological support base for the immigrant group and aids in the provision of services to that community. With the exception of certain recently arrived immigrant and refugee groups, as well as public housing tenants, there appears to be little relationship in Toronto between residential segregation and social exclusion.[17] For example, both the Jews, who are highly segregated, and the Germans, who exhibit very little residential segregation, have attained occupational and income equality with the host population (Reitz 1990). These groups are also well positioned economically because of the relatively diverse occupational structure of the Germans and the concentration of Jews in managerial and professional jobs.

The Italians and Portuguese exhibit moderately high levels of residential segregation. Male employees from both groups are heavily concentrated in the construction trades, although this is changing somewhat for the longer-established Italians who are increasingly attaining higher levels of education and moving into managerial and professional jobs. Those engaged in the building trades benefited from Toronto's emergence as a post-industrial metropolis by finding employment in the construction of new offices and residential subdivisions. Many became home owners early in the immigrant experience and profits made in real estate during the boom years of the 1970s and 1980s enabled them to move first to the inner suburbs and subsequently to the outer suburbs. Many also established small businesses, thus providing employment opportunities for others from the same ethnic group. The construction industry is cyclical and these groups, especially the older generation, may be affected negatively by both the slow-down in privately financed construction projects in the 1990s and the withdrawal of the public sector from new social housing construction. Older generation female employees from Italian and Portuguese backgrounds also face an uncertain future because they tend to be concentrated in the declining manufacturing sector and in low-paid service jobs (Giles and

Preston 1991). Some, however, have taken on paid work at home as an alternative, thereby adjusting to the new economic realities of the post-industrial city (Giles and Preston 1996).

More recently arrived groups such as the South Asians, Chinese and Blacks exhibit the same moderately high levels of residential segregation as the Italians and Portuguese. Although there is a high level of diversity within these groups, they have generally not done as well economically as the other groups because they are in declining or low-paid sectors of the economy. The future prospects of the most marginalised in these groups may also be in jeopardy, not because of residential segregation, but due to recent actions by the provincial government. Reductions in welfare expenditures, job creation programmes and day-care provision will probably have the greatest impact on the most vulnerable in society.

One of the most recent and vulnerable groups in Toronto is the refugee population. Relatively little specific information is available about the settlement experiences of this group, although a recent survey of the Somali community provides some insights (Opoku-Dapaah 1995). Almost 25,000 Somali refugees have come to Canada since 1981. The majority (80 per cent) came after 1987 and settled in Toronto. Most live in high-rise apartment buildings in the inner suburbs, often in overcrowded conditions due to the large size of many Somali families. Although indices of dissimilarity are not available, it is evident that the majority of the Somali population is spatially concentrated in both public housing and low-rent private accommodation. About 30 per cent are employed and half are on social assistance. Those who work are primarily engaged in factory jobs, retailing, low-level clerical positions and the service industry. Not unexpectedly, household incomes are low by Canadian standards. A major question is whether the Somalis are in danger of becoming a permanent 'underclass' or whether they simply suffered the misfortune of arriving in Canada at the beginning of a severe economic recession without the support of an existing community structure and with relatively little access to official resettlement services. Only the future will provide a definitive answer but Opoku-Dapaah (1995: 96) concludes that in spite of multiple disadvantages the Somalis do not constitute an 'underclass'. One reason is that in spite of the short period of time that the Somalis have been in Toronto there is evidence that language proficiency, occupational status and income have improved with length of residence.

The most likely group in Toronto to suffer permanent social exclusion are the long-term residents of public housing. In Toronto, public housing is a residual form of social housing developed to accommodate low-income households. Entry to the system is on a points basis and rents are based on tenant income. Development of public housing was terminated in the late 1970s and replaced by non-profit and co-operative housing that is designed to house a wider range of income groups. As noted earlier, much of Toronto's least well-off population still lives in parts of the newly emerging executive central city and evidence points to the increased income polarisation in that part of the city. Evidence from the 1971, 1981 and 1991 censuses also points to the emergence of the inner suburbs as the home of many of Toronto's least well-off households: low-income families, service employees, the unemployed, low-income Blacks and new immigrants. In part, this relates to housing opportunities, particularly the development of low-cost public rental housing in parts of Scarborough, North York and Etobicoke during the 1960s and 1970s when these municipalities were growing rapidly and government funding was available for public housing. Much of this housing was built on marginal sites in relatively less accessible areas of the city. In addition, low-rent private sector apartments tended to be located close by, creating mini-concentrations of the poor.

Evidence indicates a substantial growth of marginalised groups in housing owned and managed by the Metropolitan Toronto Housing Authority (MTHA), the major public housing provider in Toronto (Murdie 1994a and 1994b). In social terms MTHA developments became much more unlike the rest of the Toronto CMA between 1971 and 1986. Based on indices of over-representation, Black visible minority population showed the greatest relative differentiation (5.5 times the CMA percentage) followed by single-parent families (4.5), male unemployment (4.5) and average household income (3.9) (Figure 5.4). The concentration of these groups in MTHA housing also increased considerably between 1971 and 1986. In percentage terms, Black visible minority population increased from 4.2 per cent in 1971 to 27.4 per cent in 1986 and single-parent families went from 25.2 per cent to 41.5 per cent.

There is also considerable social variation within the public housing system. In particular, developments in 1990 with a high incidence of social deprivation (low-income, single-parent, welfare-dependent households) were mainly located in the inner suburbs, particularly Scarborough (Figure 5.5). In the ten most deprived developments, 38 per cent of households on average had very low incomes, 77 per cent were single-parent and 58 per cent were dependent on social assistance (Murdie 1994a: 100). All these figures were considerably above the MTHA average. Most of these developments are in areas of the city that are not well served by public transportation, although the residents may be closer to manufacturing and other relatively low-skilled jobs that have vacated the central city.

Because they are reserved exclusively for low-income groups, public housing areas in Toronto are those where poverty is most likely to be concentrated. These areas are also those where residents are likely to be most socially excluded and have the fewest life chances. It is difficult for residents to escape from this form of concentrated poverty. Not only is unemployment high but when people are searching for a job there is a stigma attached to an MTHA address. Public housing in Toronto has become increasingly a residual form of social rented housing, a shift noted by Harloe (1995) for various European countries where the trend has been towards accommodating an increasingly narrow segment of

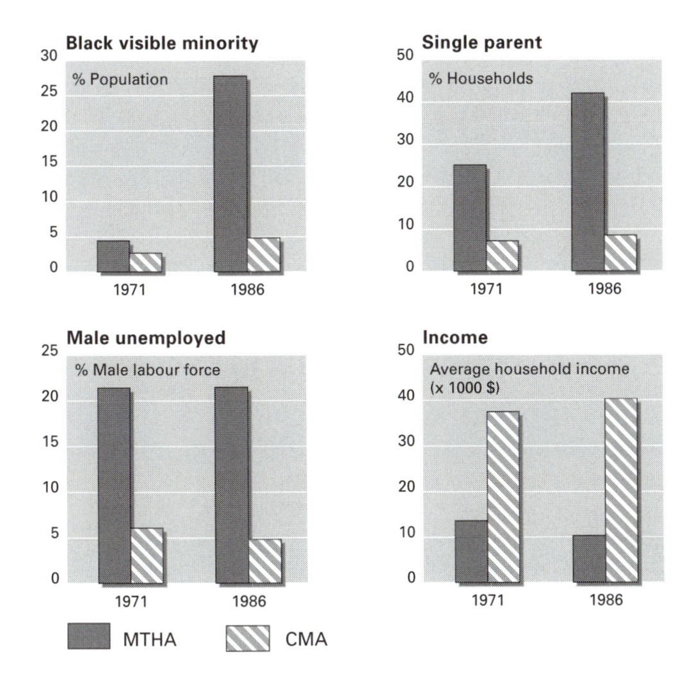

Figure 5.4 Social differentiation between Metropolitan Toronto Housing Authority and the rest of the Toronto Census Metropolitan Area

Source: Murdie (1994b: 446), Figure 2

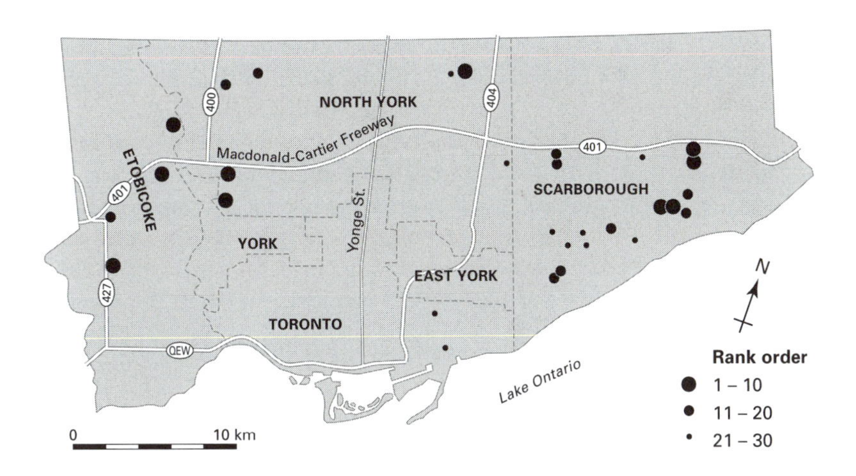

Figure 5.5 Metropolitan Toronto Housing Authority developments with a high incidence of deprivation, 1990

Source: Murdie (1992) *Social Housing in Transition: The Changing Social Composition of Public Sector Housing in Metropolitan Toronto*, Toronto: Canada Mortgage and Housing Corporation, Figure 13, p. 99

society in social housing. However, because social housing in most European countries accounts for a much larger percentage of housing stock than in Toronto, it is likely that the extent of social residualisation is not nearly as severe.[18] There is also some evidence indicating that social polarisation within MTHA stock is much more extreme than in the municipal housing component of metropolitan Sweden's Million Programme housing (Murdie and Borgegård 1992: 15). Even so, it seems that Toronto's public housing is not as physically and socially distressed as public housing in many United States cities (e.g. Dreier and Hulchanski 1993, Schill 1993). Nor are the results of concentrated poverty in public housing as extreme, although many of the same problems such as school drop-outs, marginal positions in the labour market and deviant activities exist.

Conclusion

In the last decade Toronto has experienced changes in economic structure that parallel those of many other large metropolitan centres in the western world. These include a decline in manufacturing activity, suburbanisation of much of the manufacturing activity that remains, and an expansion of producer services and financial activities, especially in the central city. Accompanying these changes, particularly during the recessionary period of the early 1990s, was an increase in unemployment and part-time jobs. During the post-Second World War period, the ethnic composition of Toronto's population also changed dramatically. Throughout this period Toronto attracted a large immigrant population but the countries of origin changed significantly following a shift in government policy in the late 1960s. This shift meant a much greater internationalisation of Toronto's immigrant population, and in recent years the inflow of immigrants and refugees from all parts of the economic spectrum.

There have also been changes in Toronto's socio-spatial landscape over the past two decades. Many of these shifts are associated with the economic and ethno-cultural changes noted above. Accompanying the increased importance of producer services has been the creation of a large number of new managerial and professional jobs, thereby supporting the professionalisation hypothesis. Many of these managerial and professional employees work in the new office space that has been created in the central city. The construction of luxury condominiums and the gentrification of existing residential areas, combined with the continued presence of low-income housing, has resulted in an increasingly polarised income structure in that part of the metropolitan area.

Although many of Toronto's most disadvantaged residents still live in the central city, analysis of 1971 and 1991 census data indicates an increased incidence of low-income households, low educational achievement, lower levels of occupational status and higher unemployment in the inner suburbs. These changes result from two phenomena: the increased number of immigrants and refugees in the inner suburbs and the construction of relatively large amounts of public housing in the 1960s and 1970s in that part of the metropolitan area. The Italians, many of whom arrived in Toronto in the 1950s and 1960s, have progressively abandoned their original area of settlement in the old core of the city and moved to the inner suburbs and increasingly the outer suburbs. More recent immigrant groups such as the Blacks are highly over-represented in the inner suburbs, partly because of the concentration of this group in the public housing developments that are located there. There has also been a shift in the location of recent immigrants during the 1971 to 1991 period from the traditional immigrant reception areas in the central city to the inner suburbs, partly as a result of the lower rent opportunities, but also because these are areas where business-class immigrants can find relatively new single detached family housing.

In general, there is considerable evidence that the immigrant population in the Toronto area has achieved a comparatively high degree of social mobility and home ownership, although there are differences between groups. There are also substantial differences in the spatial segregation of ethnic groups although there is no evidence that segregation is necessarily negative and results in a high level of social exclusion. Recently arrived refugee groups such as the Somalis are probably the most disadvantaged although it appears that these groups also have the potential for achieving social mobility within a few years in the same way as their predecessors.

As in other countries that have a strictly residualised public housing sector, occupants of public housing in Toronto are probably the most socially excluded group in the metropolitan area with the fewest life chances. Socially, public housing in Metropolitan Toronto has become increasingly residualised from the rest of the population and also differentiated within, so that some developments are much more 'socially distressed' than others. No exclusively low-income public housing has been built in Toronto since the late 1970s. Instead, emphasis shifted to the development of mixed-income social housing funded by the federal and provincial governments but sponsored by local municipalities, other non-profit groups such as churches and labour unions, and co-operatives. Continued development of this form of housing, however, has been curtailed by the federal and provincial governments. The recently elected provincial government favours a rent supplement or housing voucher option instead of a non-profit housing programme.

One of the major factors that minimised social polarisation in Toronto in the post-Second World War period was the development by senior levels of government of a modest but relatively secure set of social assistance programmes. These were not as wide-ranging or as generous as those offered by many Western European countries but they were, on the whole, more generous than the American model. In short, a status quo position was retained through the economic cycles and immigrant waves of the 1960s, 1970s and 1980s that minimised the social hardships of downturns in the business cycle and eased the economic adjustment of new immigrants and refugees. Now, however, the social welfare model is changing – abruptly and harshly. The major factors that will increasingly affect the degree of social exclusion in Toronto are the reduction in transfer payments and proposed restructuring of the social welfare system by the federal government and the substantial reductions in welfare benefit payments, reduced day-care provision and cancellation of non-profit housing programmes by the Ontario government. In the longer run, the Ontario government assumes that its plan to reduce the personal income tax rate and encourage the well-off to spend more money will create the jobs that are needed to get people off welfare and into the labour market. The poor in British and United States cities who have experienced the inequalities of similar neo-conservative policies may not be as certain.

Notes

1 The boundary of the Census Metropolitan Area also expanded slightly over this time period.

2 Toronto's index of ethnic diversity increased from 0.61 in 1961 to 0.80 in 1991 (Balakrishnan and Hou 1995: 9). The index is defined as $1-\Sigma P_i^2$, where P_i is the proportion in the *i*th ethnic group. Toronto is one of seven Canadian census metropolitan areas with indices of 0.80 or above in 1991. The other census metropolitan areas, of which Winnipeg has the highest index (0.85), are primarily in western Canada. All these cities have a long experience of immigration from various European countries.

3 Social security benefits refer to 'Compensation for loss of income for the sick and temporarily disabled; payments to the elderly, the permanently disabled and the unemployed; family, maternity and child allowances and the cost of welfare services' (United Nations Development Programme 1994: 222). Using a different data set for 1990, Oxley and Martin (1991: Table 2) place Canada seventh among the eight nations considered here in expenditure on social security and other current transfers, slightly above the United States, about even with Britain, and considerably below the remaining countries.

4 Canada's unemployment rate has been considerably higher than that of the United States since 1982.

5 In Figure 5.1, Metropolitan Toronto includes the City of Toronto, East York, York, Etobicoke, North York and Scarborough.

6 These areas approximate the outer suburbs shown in Figure 5.3. Metropolitan Toronto and the regional municipalities of Durham, York, Peel and Halton comprise the Greater Toronto Area, the boundaries of which are slightly different from those of the Census Metropolitan Area.

7 Recent provincial government legislation has made it more difficult for apartment owners to convert their buildings, particularly when private rental units are in short supply.

8 The income ratio can be viewed as an index of over- or under-representation compared to the Toronto Census Metropolitan Area (CMA). The income for the respective area is divided by income for the CMA. An index of 1.0 indicates no difference between the respective area and the CMA. The greater the distance of the index from 1.0, the greater the difference from the CMA. The index has the further advantage of standardising for temporal changes in values for the CMA.

9 The data are from the 1971 and 1991 censuses; incomes are for 1970 and 1990 respectively.

10 Enumeration areas were used rather than the larger and more frequently used census tracts in order to capture the finer grained spatial differences in income differentiation, particularly within the inner city. New luxury condominium buildings, for example, often correspond spatially with a single enumeration area.
11 Evidence reported by Bourne (1994: 567 and 573) also indicates increased income polarisation in the City of Toronto between 1970 and 1990 and a higher level of polarisation in the City of Toronto in 1990 than in the metropolitan area as a whole.
12 Low-income families are economic families whose income falls below Statistics Canada's low-income cut-offs. An economic family is a group of two or more persons related to each other and living in the same dwelling. The data for occupational status refer to the residence of employees rather than their place of work.
13 This observation corresponds with Bourne's (1994: 566) findings for average household income over a longer time period (1950–90). Between 1950 and 1990, income ratios for the City of Toronto compared to the CMA remained about the same, whereas for the five inner suburbs they declined from an average of 109.3 in 1950 to 88.3 in 1990. In contrast, for a sample of six outer suburbs, the ratios increased from an average of 110.9 in 1960 to 121.9 in 1990.
14 Ethnic origin refers to the cultural background of the respondent. The data for 1981 and 1991 are for single origin groups only. In the 1971 census only one ethnic origin was reported per respondent. In the 1981 and 1991 censuses respondents were allowed to report multiple origins.
15 These figures exclude the refugee population, who probably earn considerably less than immigrants. A recent study by Opoku-Dapaah (1995) of a sample of Somali refugees in Toronto found that only 8 per cent of households in this group had incomes above $30,000. The median household income for all immigrants who lived in the Toronto Census Metropolitan Area and came into the country in 1986–91 was $33,500 (Lapointe and Murdie 1995: Table A6).
16 The income figures have not been adjusted for other factors such as education, age of household maintainer or household type. For Canada as a whole, Balakrishnan and Hou (1995: 31) have demonstrated, using individual wage income for 1990 and controlling for a variety of other factors, that visible minority groups such as the Chinese, South Asians, Central Americans and Blacks/Caribbeans have wage incomes that are upwards of $6,000 less than the national average of about $29,000. These relative discrepancies did not improve between 1980 and 1990.
17 A relatively large number of recently arrived immigrants are housed in public housing although, because of long waiting lists, not all who apply can obtain accommodation.
18 Specific comparison across countries is difficult, as Goering (1992) found in his attempt to compare public housing segregation between the United States and England.

6

EXCLUSION AND INCLUSION

Segregation and deprivation in Belfast

Frederick W. Boal

Social exclusion

The central theme of this volume focuses on segregation and exclusion in western metropolitan areas. Such a focus immediately brings to mind the 1993 Green Paper issued by the European Commission, the objective of which was to open up discussion on options for the future of European social policy (Commission of the European Communities 1993). In this document the Commission stressed the need to tackle social exclusion, where emphasis

> is now on the structural nature of a process which excludes part of the population from economic and social opportunities. The problem is not only one of disparities between the top and the bottom of the social scale, but also between those who have a place in society and those who are excluded.
>
> (Commission of the European Communities 1993: 20)

The Commission further noted that exclusion affects not only individuals who have suffered serious setbacks, but social groups who are subject to discrimination, segregation or the weakening of the traditional forms of social relations. Finally, it was noted that the causes of exclusion are multiple – persistent unemployment, the impact of industrial change on poorly skilled workers, the evolution of family structures and the emergence of new forms of migration (in particular illegal immigration).

Closure, exclusion and usurpation

While broad structural forces lie at the root of exclusion, many situations also arise where certain groups suffer from social exclusion at the hands of other groups – in other words, there are not only the excluded but the excluders. To develop this idea further and subsequently to link it to the question of residential segregation, it is useful to refer to the concept of social closure, as first developed by Max Weber (1968) and as refined subsequently by Frank Parkin (1979). By social closure Weber meant the process by which social collectivities seek to maximise rewards by restricting access to resources and opportunities to a limited circle of eligibles (Parkin 1979: 44). This is achieved by singling out certain social or physical attributes as criteria to designate those to be excluded – race, ethnicity, language, religion, social origin, etc., will serve, separately or in some combination. Groups designated in this way become outsiders and are subject to closure in terms of social and economic opportunities.

Parkin takes the concept of social closure and develops it in such a way as to increase its subtlety and sophistication. He suggests that closure strategies should not just include those of an exclusionary kind, but also those adopted by the excluded themselves as a direct response to their status as outsiders. This latter form of closure he designates 'usurpation'. Exclusionary closure represents the use of power in a downward direction because 'it necessarily entails the creation of a group, class or stratum of legally defined inferiors' (Parkin 1979: 45). What Parkin calls countervailing action by the negatively privileged, on the other hand, represents the use of power in an upward direction as the excluded seek to win a greater share of resources by biting into the privileges of those in superior socio-economic positions.

Exclusionary closure and usurpationary closure form the two basic strands of Parkin's formulation. He goes one step further, however, by suggesting that exclusionary closure is not just something perpetrated by a superior class on an 'inferior' one – it is a process that frequently occurs within the subordinate class itself as one segment attempts to gain advantage by undertaking exclusionary closure at the expense of even more vulnerable groups, such as ethnic minorities or women (Parkin 1979: 91, Kilmurray 1995). Thus we may find groups in society that suffer from exclusionary closure while at the same time they themselves attempt to respond by upwardly directed actions of usurpationary closure and by downwardly directed actions of an exclusionary nature.

The functions of segregation

Segregation tends to have negative, exclusionary connotations. Certainly, if it is forced on a given group, then its members are unlikely to view their situation with much favour. On the other hand, it is important to recognise that segregation can have positive dimensions, particularly when groups self-segregate from choice, less so where segregation is a group's response to a conflict-ridden environment. What are these positive functional attributes?

First, segregated areas can provide arenas for the maintenance of particular ways of life, whether these be defined in broadly cultural or more narrowly religious terms (Boal 1987). Second, segregated population concentrations offer environments supportive of ethnic entrepreneurship. Third, segregation can provide an organisational basis for action in the wider society – focused on institutional politics or, in certain circumstances, on insurrectional activity. Some parallels to these situations may be found in the situation described by Wacquant and Wilson (1993) when they make reference to what they call the 'organised' or 'institutional' Black ghetto of mid-twentieth-century American cities, where 'activities are . . . structured around an internal and relatively autonomous social space that duplicates the institutional structure of the larger society and provides basic minimal resources for social mobility' (Wacquant and Wilson 1993: 32). Segregation may also be viewed, though only in response to unfavourable circumstances, as having a positive function where members of particular groups fear attack; here physical security can be increased by residing in an area of relatively homogeneous ethnicity. Likewise, minority groups may wish to practise avoidance – that is, they may wish to distance themselves from excessive contact with dominant population groups, thus keeping clear of potentially embarrassing or psychologically threatening situations, where their personal appearance, limited majority language ability or aspects of their lifestyle may be subjected to abuse or ridicule.

Ethnic segregation, therefore, may be highly functional for the group or groups so segregated. If, however, the segregation is basically involuntary, then the segregated space may be less of an expressive or instrumental resource and rather more of a trap. Indeed, such segregation may be highly functional, not for the ethnic group or groups concerned but more for the wider society. Thus the defensive function may be seen as one of containment; avoidance may meet the prejudicial needs of the majority community, enabling them to avoid sustained contact with 'strange' ways of life. Environments that contribute to the maintenance of ethnic cultures may also contribute to the protection of majority culture from 'dilution' ('pollution'?), while external action may mean the manipulation of electoral geography so as to prevent or minimise the election of ethnic minority candidates. In the worst case scenario, segregation may mean the availability of clearly defined target population concentrations. It may be further argued that segregated residential contexts can serve as labour reservoirs, to be drawn upon when a large volume of low-cost labour is in demand, and to function as dumping grounds for the unemployed during sustained periods of economic downturn. Sweeney (1994) has referred to such neighbourhoods as 'economic refugee camps'.

Finally, it could be argued that where a number of ethnic groups are present, encouragement of segregation *between* the various groups provides a basis for the creation and reproduction of ethnically/racially distinct fractions of the labour force, thereby weakening the labour market bargaining position of each group separately and, indeed, of the ethnic minority population as a whole.

In all this we can see the operation of closure – exclusionary closure applied to distinctive ethnic groups by the wider society, exclusionary closure by one or more of the ethnic groups themselves applied to other ethnics and, perhaps, usurpationary closure as one or more of the disadvantaged groups strive to improve their circumstances.

When we review the pros and cons of residential segregation it is clear, as Stephen Castles claims, that it is a contradictory phenomenon. Writing about migrants he concludes that

> [they] may be socially disadvantaged by concentration in areas with poor housing and social amenities but they frequently want to be together in order to enjoy mutual support, rebuild family and neighbourhood networks, and maintain their languages and cultures. Ethnic neighbourhoods allow the establishment of small businesses and agencies which cater to migrants' needs, as well as the formation of associations of all kinds. Residential segregation is thus pre-condition for and a result of community formation.
>
> (Castles 1993)

And yet such a relatively rosy picture may seriously understate the negative aspects of segregation. Segregation may mean exclusion from resources such as housing of acceptable quality or job-rich social networks. Disadvantage may be reinforced by spatial concentration and a concomitant 'culture of poverty' – to use Wacquant and Wilson's terminology the 'organised ghetto' may become the 'disorganised ghetto'. In such circumstances selective out-migration may further accelerate processes of deprivation as those with opportunities for spatial and social mobility leave, producing a highly residualised population stranded in the segregated neighbourhoods. To add insult to injury, such residualised neighbourhoods are frequently subjected to negative stereotyping, a stereotyping that is all too easily transferred to the inhabitants themselves.

Belfast

Belfast saw its beginnings as a fortified crossing point of the River Lagan. A castle was constructed as early as the 1170s, but even by the eighteenth century Belfast was still a modest place, having a population of only 20,000 by 1800. The nineteenth century saw a great transformation, however, as the city grew rapidly under the stimulus of an industrialisation that was primarily based on linen manufacture, shipbuilding and general engineering. By the beginning of the twentieth century population numbers had reached 350,000. This figure was achieved by natural increase founded on major in-migrant flows from much of the northern part of Ireland. These migrant flows introduced a significant Catholic component into what previously had been an overwhelmingly Protestant town. Catholic–Protestant friction grew with this expansion. As has been pointed out by Maguire (1993), an influx of Irish Catholic labour was a common feature of nineteenth-century industrialisation not only in Belfast but also in cities in Britain such as Glasgow, Liverpool and Manchester. All had their inter-ethnic tensions, but in the case of Belfast these were to prove in every way sharper, more persistent and more divisive than anywhere else.

Evocatively, Maguire (1993) describes this movement as 'Catholic counter-colonisation', importing with it intense feelings of territoriality, common enough at the time for both Protestants and Catholics in rural areas to the west.

The latter part of the nineteenth century also saw the emergence of a powerful Irish nationalist movement that sought some degree of independence from the English-dominated United Kingdom. This was strongly opposed by the Protestant population in the north of Ireland, nowhere more so than in Belfast itself. In consequence, what had been an ethnic division in the city became an ethno-national one. The partition of Ireland in 1920 further intensified ethno-national conflict, a conflict where the labels distinguishing the groups concerned are religious (Catholic/Protestant), but where difference is not limited to theology but is rooted in different senses of origin, different senses of history and differences in perceived national identity (Smith 1986). In sum, all these ethnic differences are socially marked by religion. At the same time it must be stressed that religion is not merely a social marker – it provides a basis for differences in attitudes and behaviour that themselves contribute significantly to the overall ethnic differentiation (Fulton 1995, Boal *et al.* 1996).

The first half of the twentieth century saw Belfast continue to grow, though more slowly than during the dynamic latter part of the nineteenth. By mid-century, indeed, population expansion in the urban core had ceased, though suburban growth continued until 1971. The traditional employment base of the city was sustained until the end of the 1950s. Thereafter, however, there was a radical transformation as manufacturing employment declined and service-related employment increased. Thus, in 1961 employment was split 50:50 between manufacturing and services; by 1993 manufacturing had declined to a point where it comprised only 17 per cent of overall employment (Figure 6.1). This employment restructuring, profoundly important in its own right, has also had a substantial differential impact on the two major ethnic communities in the city.

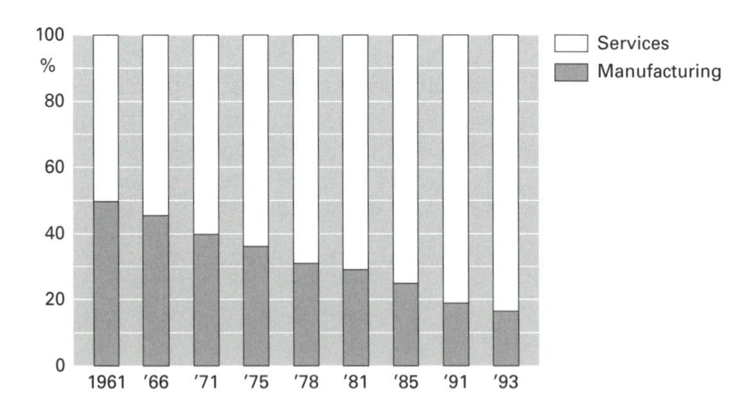

Figure 6.1 The structure of employment in the Belfast Urban Area, 1961–93

While employment patterns were being transformed, so also population distribution was changing quite radically. To understand what happened in this regard it is useful to recognise a number of components of Belfast – an inner core that spatially corresponds with the non-suburban section of the nineteenth-century industrial city, the suburban fringes of the city as it existed up to the First World War, the suburban zone of the present urbanised area and, finally, a commuting zone of discontinuous urban settlement that can be referred to as the outer regional city. If we look at what has happened between 1971 and 1991 we find that the population in the inner core has declined by 55 per cent, while there has been a decrease of 22 per cent in the older suburbs. On the other hand, the newer suburbs have grown by 8 per cent and the outer regional city by 39 per cent (Boal 1995). Obviously, then, there has been a fundamental spatial redistribution of population, with the most profound effects concentrated in the inner city (Figure 6.2), an area we will examine more fully when we turn to discuss deprivation and social exclusion.

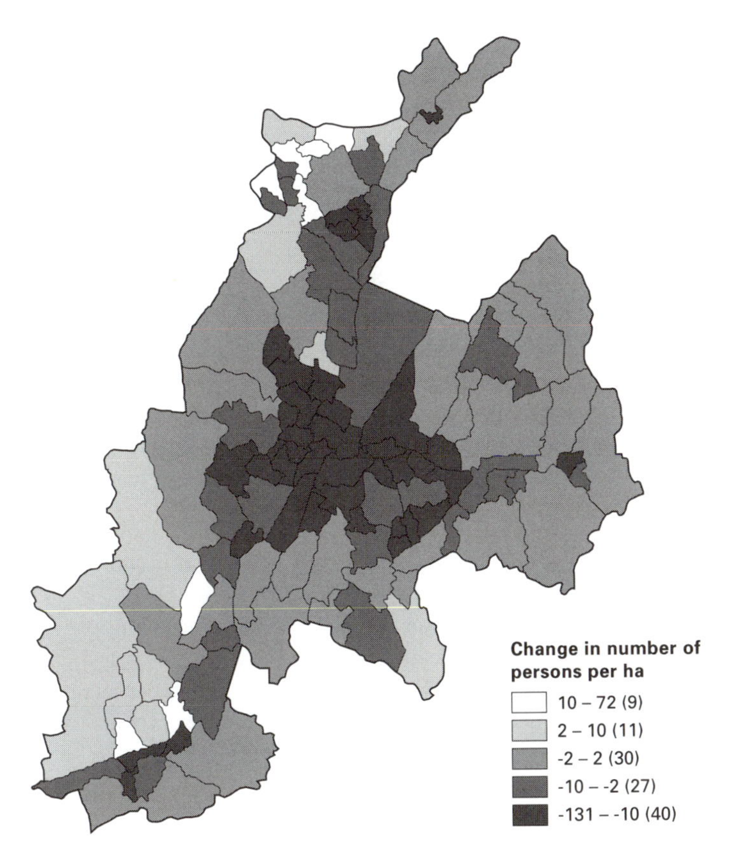

Figure 6.2 Belfast Urban Area: increase and decrease in population, 1971–91

One final point regarding recent population movement needs to be made, particularly in the context of a volume where most of the cities being examined have recently seen major flows of international immigrants. This has not been the case with Belfast. There are small communities of Chinese and of Indians, but a declining employment base with concomitant high levels of unemployment, combined with a quarter of a century of violent conflict involving the two 'native' ethnic groups, has not been conducive to immigration. Even if we consider the Catholic and Protestant populations themselves as deriving ultimately from in-migrant/immigrant stock, historical circumstances have determined that assimilation processes that might normally have eroded, or, indeed, removed ethnic difference, have not, in fact, taken place – migrant-based relationships have, as it were, become frozen in time.

Ethnic segregation

Catholic–Protestant conflict in Belfast has already been alluded to. Perhaps the most distinctive manifestation of this conflict has been the persistence, indeed the strengthening, of residential segregation between the two populations (Figure 6.3). Current levels of segregation, undoubtedly higher than they have ever been in the past, have been arrived at by what may well be seen as a distinctive process. This process can be designated the segregation 'ratchet' (Figure 6.4). Here we see that inter-ethnic segregation levels have risen considerably over the past 150 years. This rise has not been a steady one, however. Rather it has occurred as a series of jolts, each jolt followed by relatively lengthy periods of segregation stability, or even by some decline in segregation severity (see Doherty and Poole 1995). The sharp increases corresponded with periods of communal instability. Most significantly, however, intervening, more tranquil periods do not display anything more than small decreases in segregation. Consequently the next outburst of inter-communal violence takes off, as it were, from the platform of segregation installed in the previous violent episode. While available evidence indicates that Catholic–Protestant segregation in Belfast before the renewed violence that commenced in 1969 was less than White–Black segregation in the United States (Poole and Boal 1973), by the late 1970s segregation levels had converged considerably, due to the combined effect of increase in Belfast and some decrease in urban America.

The second aspect of segregation in Belfast that we need to note is the situation in the public (social) housing sector. Here, immediately before 1969, ethnic segregation was close to the overall level in the urban area as a whole (a Dissimilarity Index of 69.4).[1] However, by 1977 the recorded index in the public sector had leapt to 91.8 (Keane 1990). This is high even by American standards and clearly indicates that the public housing sector carried the brunt of the escalating segregation that took place in the early years of the current conflict. The fact that ethnic segregation is at its sharpest in that sector of housing provision that should be most amenable to central control may seem

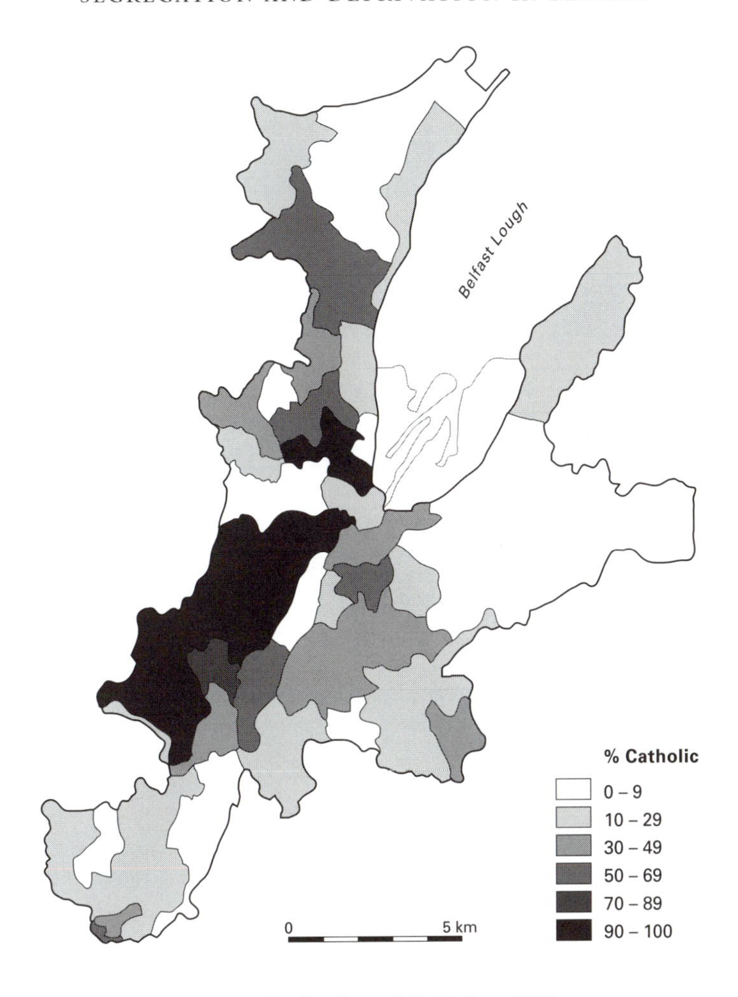

Figure 6.3 Belfast Urban Area: distribution of Catholics, 1991

surprising. However, the circumstances that prevailed in Belfast in the 1970s created conditions highly unconducive to ethnic residential mixing. There was great insecurity and many households were displaced from mixed and from 'frontier' environments. The public housing authority had a statutory obligation to rehouse such households, who, in the very nature of their circumstances, were likely to seek the safety of accommodation in neighbourhoods dominated by their own group. An additional factor increasing segregation in the social housing sector was that many inner-city housing areas, already highly segregated, went through a tenure switch as part of the redevelopment process. In this instance dwellings previously rented from private landlords, or in some instances owned by their occupants, moved into public ownership and were subsequently replaced by public sector dwellings. It should be emphasised that

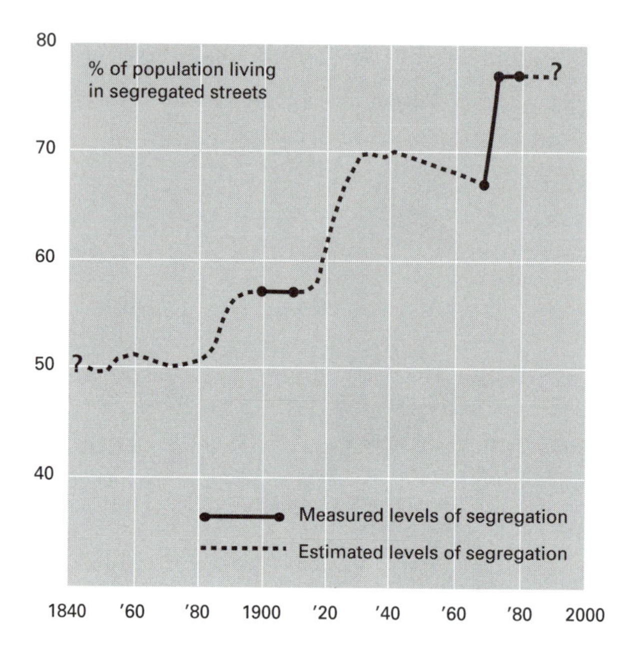

Figure 6.4 The Belfast ethnic segregation ratchet (segregated streets are defined as those where the population is over 90 per cent Catholic or over 90 per cent Protestant)

the public housing authority (the Northern Ireland Housing Executive) favoured ethnic mixing in principle, but felt that it would be utterly wrong to attempt to apply such an approach to housing allocation in what was a highly stressful, conflict-ridden environment (Northern Ireland Housing Executive 1972).

While ethnic segregation, *per se*, is a dominant theme in Belfast, it co-exists with segregation based on social class/income differentials. In Belfast ethnic segregation basically cross-cuts ethnic segregation, rather than corresponding with it. In other words, members of both ethnic groups are found in all the social strata, though historically there has been some differentiation, in that Catholics were disproportionately represented in the lower strata, Protestants in the higher. When we examine the cross-cut of ethnic and social class segregation we find that the latter is almost as high as the former. Most importantly we find that the lower-income segments of both the Catholic and the Protestant communities are quite sharply segregated from the higher-income segments of their own group – that is, intra-ethnic segregation on the basis of class is a powerful dimension of the overall residential differentiation. Beyond this we may also note that lower-income Catholics are more highly segregated from lower-income Protestants than higher-income Catholics are segregated from higher-income Protestants (Figure 6.5).

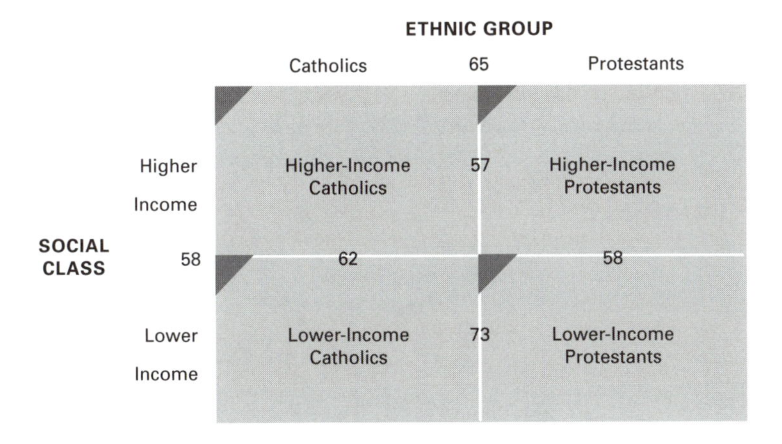

Figure 6.5 Belfast Urban Area: ethnic and class segregation, 1973

From this somewhat complex analysis one highly salient feature emerges: namely that the two most severely segregated segments of Belfast's population are the low-income Catholics and the low-income Protestants. These Catholic and Protestant groupings are class-segregated from their 'own' middle classes and they are ethnically segregated from each other. This situation emerges as a salient dimension when we turn to examine aspects of deprivation.

Deprivation

As with many other cities, the spatial distribution of deprivation has been mapped for Belfast. The first of these exercises, using data from the early 1970s, was carried out by the Geography Department at the Queen's University of Belfast (Boal, Doherty and Pringle 1974), showing a concentration of 'problems' in the inner city, in a sector extending to the west and in a number of outlying suburban pockets. This analysis was subsequently extended and updated by a government team (Project Team 1977). The latter investigation had the advantage of access to a more extensive range of data sources, thus permitting a more broadly based assessment of need. A range of indicators was used – unemployment rates, ill-health, housing quality, mental disability and juvenile delinquency amongst others. Not surprisingly, the geography of deprivation disclosed (Figure 6.6) varied little from that recorded in the earlier exercise. The same deprivation is seen in the maps provided by Murtagh (1994) – using data from the mid-1980s – and in the analysis carried out by Manchester University which used 1991 census material as the predominant source of indicators (Robson, Bradford and Deas 1994) (Figure 6.7). The 1991 mapping was based on fourteen indicators of deprivation, broadly in line with the 1977 exercise. All these mappings were carried out to provide criteria for targeting aid at the most deprived sections of the city, and have, indeed, been thus employed – for instance in the 1970s Belfast

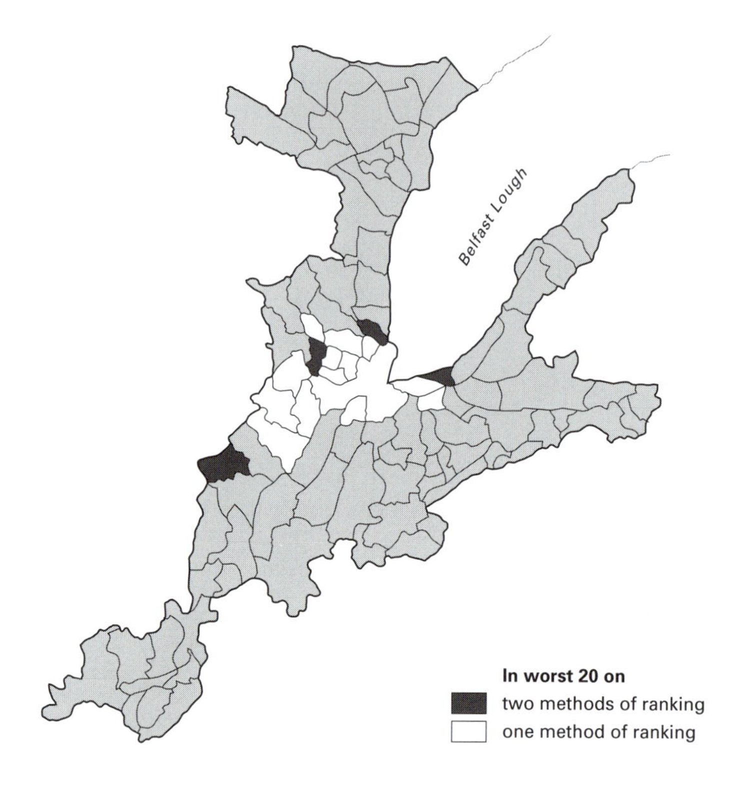

Figure 6.6 Belfast Urban Area: areas of need, 1976

Areas of Need (BAN) strategy and in the current Making Belfast Work (MBW) programme (Making Belfast Work 1995).

While housing quality has been radically transformed for the better over the past two decades, the fundamental locational stability of deprivation emerges as a striking feature. The aid programmes may have helped ameliorate disadvantage, but it still persists with a depressing degree of spatial and social stubbornness. This makes me uneasily aware of David Harvey's 1973 admonition that 'Mapping even more evidence of man's inhumanity to man is counter-revolutionary in the sense that it allows the bleeding heart liberal in us to pretend we are contributing to a solution when in fact we are not' (Harvey 1973). When we turn to examine the relationships between deprivation and residential segregation, we find, not in the least surprisingly, that there is a close correspondence with social class. It is the working-class areas that take the brunt of disadvantage. We also find that there is a related correspondence between deprivation and the areas of highest *ethnic* segregation. Moreover, in all this the public sector has a striking role, in that the most deprived areas have a high proportion of their housing stock in public ownership. The quality

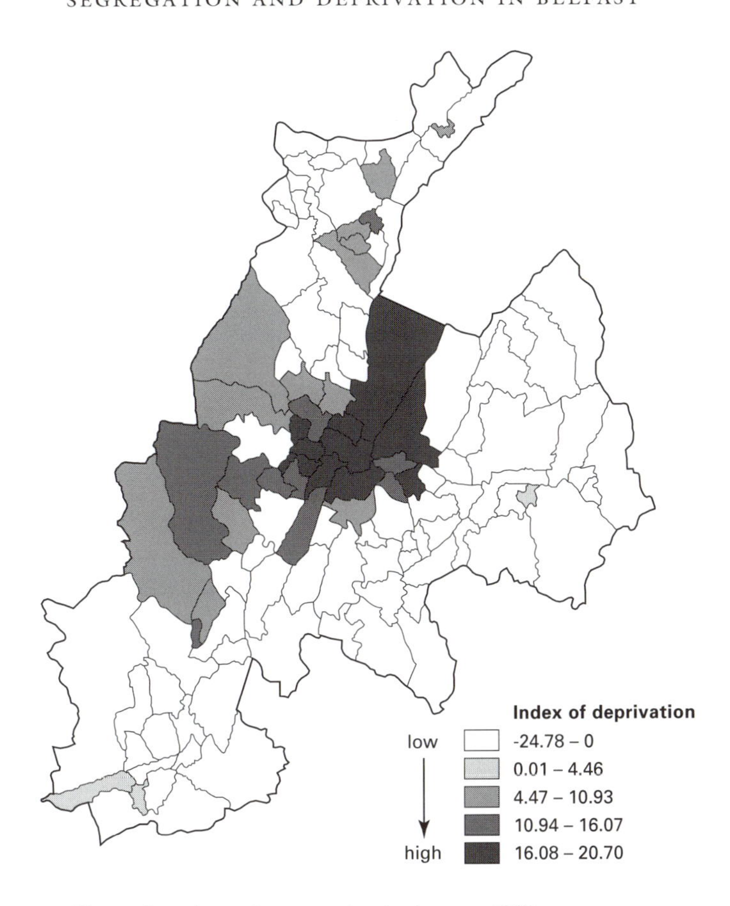

Figure 6.7 Belfast Urban Area: the most deprived areas, 1991

of this stock, because of the massive redevelopment and improvement programmes in the 1980s, is, with few exceptions, very high indeed (Northern Ireland Housing Executive 1991). At the same time, as already noted, ethnic segregation is also at its highest in this same tenure. Thus, in the 1990s, we have a good physical fabric acting as container for a population with considerable social deprivation, that population, in turn, being segmented into two highly segregated ethnic components.

Temporal change in deprivation

We can now turn to an examination of the dynamics of this situation. Here we can draw on the work of Power and Shuttleworth (1995), who have carried out a temporal analysis of change between the population censuses of 1971 and 1991. They have developed a nine-category regionalisation based on the wards in the Belfast Urban Area, distinguishing the wards on the basis of their

Table 6.1 Belfast Urban Area unemployment percentages, 1971 and 1991

Ward typology	*1971*		*1991*	
	Catholic	*Protestant*	*Catholic*	*Protestant*
Type 1	5.3	2.4	6.2	5.4
Type 2	6.9	3.1	13.2	12.2
Type 3	8.8	4.9	12.2	11.4
Type 4	12.9	8.8	18.7	23.8
Type 5	9.3	5.3	17.5	16.3
Type 6	4.4	2.4	6.5	7.5
Type 7	8.8	3.1	29.5	29.2
Type 8	17.5	9.9	34.1	31.9
Type 9	18.9	10.4	34.8	30.1
Urban area average	13.6	5.8	23.3	12.7

Source: Power and Shuttleworth (1995)

Type 1: Above average affluence Protestant wards with population increase.
Type 2: Average affluence Protestant wards with population increase.
Type 3: Average affluence Protestant wards with population decrease.
Type 4: Below average affluence Protestant wards with population decrease.
Type 5: Average affluence mixed wards with Catholic increase.
Type 6: Above average affluence mixed wards with Catholic increase.
Type 7: Below average affluence Catholic wards with population increase.
Type 8: Below average affluence Catholic wards with population decrease.
Type 9: Inner-city internally segregated very low affluence wards with population decrease.

religious composition and degree of affluence in 1991 and on the direction of population numerical change over the two decades 1971–91.

Table 6.1 shows the Catholic and Protestant unemployment rates for the various ward types for two points in time – 1971 and 1991. The *low-affluence* wards are notable for their higher rates of unemployment both in 1971 and in 1991. What is also evident is that, in 1971, within all ward types, Catholics were more likely to be unemployed than Protestants. By 1991, however, a number of dramatic changes are observable. First, overall unemployment rates have risen markedly. Second, within each ward type, Catholic–Protestant unemployment differentials have decreased considerably, and third, within the Urban Area as a whole, there is still a significant Catholic–Protestant unemployment differential, but a smaller one than in 1971 (2.3:1 in 1971; 1.83:1 in 1991). This seeming anomaly between very similar Catholic and Protestant unemployment levels in each ward category, on the one hand, and a continued Catholic–Protestant differential within the Urban Area as a whole, on the other, is largely explained by the fact that Catholics are disproportionately concentrated in a few ward types where unemployment is high – for instance type 8 wards, with a 1991 unemployment rate for Catholics of 34.1 per cent contain over one-third of all Catholics in the Urban Area (Northern Ireland Registrar General 1992).

Finally, beyond ethnic differentials, there is evidence, for the 1971–91 period, of the emergence of an increasing unemployment gap between the high- and the low-affluence wards. For Catholics, in 1971, the ratio of the unemployment rate between the highest and the lowest wards was 4.3:1; by 1991 this had opened to 5.61:1. Likewise for Protestants – the 1971 differential was 4.3:1, the 1991 differential 5.9:1.

It will be useful, at this point, to summarise the rather complex web of deprivation differentials (as measured by the unemployment rates). First, for both 1971 and 1991, there is a marked spatial variation from one part of the Urban Area to another. Second, as a group, Catholics suffer a greater degree of deprivation than Protestants. Third, the Catholic–Protestant deprivation differential has markedly reduced between 1971 and 1991, in an overall environment where deprivation has intensified. Here one might speak of an increasing ethnic equality of misery. Finally, for both Catholics and Protestants, the gap between the affluent and the disadvantaged has grown over the two decades being reviewed. Put another way, class polarisation has sharpened, irrespective of ethnicity.

Discussion

What we observe, then, is a deprivation increasingly concentrated on the highly segregated, lower-income segments of the Catholic and Protestant communities. However, while structural processes at work are broadly similar for both ethnic groups, important differences need to be recorded. First, there has been differential out-migration from the two deprived inner-city segments. Because of spatial constraints on Catholic housing choice in the Belfast Urban Area, housing densities for Catholics have declined to a considerably lesser degree than for Protestants. This has led to some overcrowding in Catholic areas, but it has also meant that 'critical mass' has been maintained in those areas to a significantly greater degree than has been the case in the Protestant inner city. Consequently, institutional support bases have been retained in Catholic areas with, for instance, community-destructive loss of schools through closure being avoided. This has also been the case with churches, though, in this instance, a higher degree of secularisation amongst working-class Protestants has made its contribution as well. Differential migration has not just affected population numbers – it has also modified population composition, with more residentially mobile young Protestants taking up suburban opportunities, leaving behind an increasingly elderly population. The lower mobility from many inner-city Catholic areas, due, as noted earlier, to more restricted choice, has helped to retain a more balanced age structure. Finally, the wider housing choice open to Protestants has also created a number of outlying working-class housing complexes which lack viable community critical mass. Adding insult to injury is the fact that a number of these small pockets of deprivation do not get picked up in social needs surveys because they get included statistically with more affluent surrounding neighbourhoods (Community Training and Research Services 1993). Thus a process of residualisation appears to be under way, a process less marked in Catholic areas for the reasons noted above. And yet, increasing social mobility amongst Catholics, due, in part, to vigorous fair employment legislation, may be leading to selective out-migration from the highly segregated Catholic areas as well, with the attendant possibility of a residualising effect similar to that already observable in a considerable number of inner-city Protestant neighbourhoods.

A second distinction between the Protestant and the Catholic deprived areas emerges from differences in 'culture'. Many community workers have testified to the relative ease with which community action can be stimulated in Catholic areas; this may be attributed, at least in part, to the historical experience of solidarity in adversity, a collective response to social problems and the Catholic church's sense of parish community. Further, the church is thought to have encouraged its politically alienated flock to develop a quasi-state under its wing and patronage – credit unions and, in the past, extensive poor relief systems are good examples. The Protestant tradition, on the other hand, adheres to different values – emphasising the individual and his or her relationship with God. The variety of Protestant denominations – contrasted with the Catholic religious monolith – serves further to diminish the basis, in Protestant neighbourhoods, for strong collective bonds to be developed (Oliver 1992). Again, the greater degree of secularisation among working-class Protestants serves to weaken the potential for church–community mutual support linkages. Furthermore, Catholic areas, by and large, have been more vigorous in their pursuit of available financial support for local community regeneration projects – whether this support comes from the European Union, the International Fund for Ireland or the many UK initiatives. In addition, it would appear that the Catholic deprived areas, up to recently at any rate, are more highly politicised than their Protestant equivalents. This has to be a subjective assessment, but, put briefly, Catholic politics derive from motives of resistance and attack, Protestant more from motives of preservation and defence. It is obvious which context is likely to generate the greater political vigour. Finally, it is worth noting that changes in the nature of employment have tended to work to the disadvantage of Protestants, who, historically, had a fairly firm exclusionary grip on employment in a number of male-dominated 'heavy' manufacturing industries, such as shipbuilding and engineering. Consequently the marked decline in manufacturing has undermined this Protestant enclosure, while the burgeoning service sector has been a much more egalitarian job source. In combination, these two trends have done much to erode the working-class Protestant employment advantage that previously existed.

Exclusion, inclusion and segregation

What conclusions can be reached from this discussion? First, and most obviously, structural and exclusionary closure processes have combined to create and maintain a disadvantaged segment of society. Second, in the case of Belfast, the disadvantaged class has been fractured along ethnic lines, a fracture where one fragment (the Protestant) has historically attempted to gain some advantage by operating a degree of exclusionary closure against the other (the Catholic). Third, we find that segregation, while the product of exclusionary closure, provides the basis for efforts to cope with such exclusion – solidarities have been nurtured, and, particularly within the Catholic community, usurpationary actions have been taken, employing a potent cocktail comprising political pressures, the politics of protest and, indeed, violence. Fourth, we have observed that the dynamics of employment restructuring and population redistribution have affected both the Protestant and the Catholic disadvantaged. However, the severest negative effects have been experienced by the former, as their community solidarities have suffered erosion and many of their historically secure job enclosures have been economically decimated. Finally, it is important to note that there has been a major convergence between the two disadvantaged groups in terms of deprivation levels – they now basically experience a shared misery that is considerably more severe (using unemployment as the key measure) than it was a quarter of a century ago.

Although ethnic and class segregation, on the one hand, and deprivation, on the other, co-exist, we must be careful not to conclude that segregation is always a negative attribute. Spatially concentrated disadvantage can, and does, create an environment where deprivation fuels the dynamics of a downward spiral (deprivation begets more deprivation). However, as has been demonstrated in this chapter, segregation can also, under the right conditions, provide a basis for coping with deprivation. Concentrated disadvantage can also expedite the application of aid programmes, programmes that derive from what is left of the UK welfare state and from European Union initiatives aimed at reducing social exclusion. The most striking impact of these policies in Belfast has been the physical transformation in the quality of inner-city dwelling stock.

One final point. Labelling highly segregated, deprived areas as 'ghettos' produces a crude analysis, and one, therefore, to be strenuously avoided.

Note

1 The formula for the Index of Dissimilarity (D) is:

$$D = 0.5 \sum_{i=1}^{n} | (x_i/X) - (y_i/Y) | * 100$$

Where x_i = the minority population in subarea i
y_i = the majority population in subarea i
$X = \Sigma x_i$ = the city-wide minority population
$Y = \Sigma y_i$ = the city-wide majority population

7

SEGREGATION, EXCLUSION AND HOUSING IN THE DIVIDED CITY

Alan Murie

The debate about changing cities in recent years has given considerable attention to global pressures for change and to the different regimes of regulation which mediate these pressures and determine the pace and extent of changes. In the advanced capitalist economies attention has also focused on how rising unemployment, labour market change, welfare restructuring and increasing social inequality have affected the population of cities. Discussion of divided and polarised cities has been informed by empirical study but most of this has been carried out at a whole-city level and issues of segregation and spatial change have often been inferred from such data.

This chapter contributes to this debate in three ways. First, it seeks to add to perspectives on the welfare state and the framework used to distinguish between different welfare state regimes in considering how the outcomes of the processes of globalisation will differ between particular countries and cities. Second, it seeks to reflect on the relative neglect of spatial patterns within cities and of segregation in a debate dominated by city-level observations. Third, it introduces material relating to aspects of the housing market in Scotland's capital city, Edinburgh.

Welfare states

Differences between cities in terms of their built environment and social and economic structure are the product of a range of factors. Among these influences is the operation and structure of the welfare state. If we ask why cities in the USA differ from those in Sweden or France, and why they continue to change in different ways, one of the contributory elements in an explanation will relate to the nature of the welfare state and its impact on patterns of production and consumption and effective demand through the market. It has been argued that because of differences in welfare state arrangements global economic pressures do not have the same effect in changing patterns and problems in cities. Thus it is argued that differences in the degree of social segregation between European and American cities and between Dutch and British cities relate to different housing and welfare state systems (Murie 1993a, Murie and Musterd 1996). Differences in these arrangements limit convergence and mediate global pressures so that local outcomes differ significantly. In this debate there has been only limited discussion of what the critical features of welfare states are in relation to urban change. The tendency has been to refer to Esping-Andersen's stimulating and coherent account of welfare state regimes and the different types of welfare states he identifies (Esping-Andersen 1990). At one level the reference to different regimes serves the essential purpose of indicating that the impacts of change will not be the same everywhere. However, at another level, the implication is that emerging patterns of change will be similar in countries falling into the same welfare regime type. This puts a considerable onus on the typology not only to raise questions and stimulate analysis but to reflect accurately the key distinctions between countries in terms of welfare state regimes. If we are to understand differences emerging in different countries, it becomes important to adopt a more reflective and critical view of welfare states. In order to address issues of urban change we need to be satisfied that the regimes of welfare identified do encapsulate how welfare states affect cities. This was not the purpose of Esping-Andersen's analysis, which has its limitations when used in analysis of urban change.

Esping-Andersen identifies three key categories of welfare state based on an analysis of social insurance and income maintenance arrangements. He refers to rules governing access to benefits, levels of income replacement and the range of alternatives and develops quantitative measures of these from which he derives a categorisation of welfare state regimes. He suggests three types of welfare state regime within advanced capitalist economies:

- Liberal welfare states associated with means-tested assistance for low-income households, the encouragement of private welfare and minimal decommodification and social rights. The USA, Canada and Australia are presented as typical.
- Corporate regimes involving the preservation of class and status and social rights which are acknowledged but attached to class and status. Extensive state provision exists but is not redistributive. Germany, Austria, Italy and France are presented as typical.
- Social democratic regimes associated with universal high-standard state services based on universal social rights, redistribution and graduated benefits and the fusion of welfare and work. Sweden and Denmark are presented as typical.

Esping-Andersen's general discussion of the modern welfare state places considerable emphasis on decommodification.

> In pre-capitalist societies, few workers were properly commodities in the sense that their survival was contingent upon the sale of their labour power. It is as markets become universal and hegemonic that the welfare of individuals comes to depend entirely on the cash nexus. . . . In turn, the introduction of modern social rights implies a loosening of the pure commodity status. Decommodification occurs when a service is rendered as a matter of right and when a person can maintain a livelihood without reliance on the market.
>
> (1990: 21–2)

There are a number of ways in which citizens are able to meet their needs other than through the market. These may involve family or community support, social insurance or means-tested social assistance. Esping-Andersen argues that insurance and assistance arrangements tend to secure only limited decommodification as entitlements relate to position in the labour market. The Beveridge-type system may be more decommodifying because it offers a basic, equal benefit to all irrespective of prior earnings, contributions or performance but only rarely have such schemes been able to offer benefits of such a standard that they provide recipients with a genuine alternative to working. The concept of decommodification is not absolute but 'refers to the degree to which individuals, or families, can uphold a socially acceptable standard of living independently of market participation' (Esping-Andersen 1990: 37). Esping-Andersen's method of operationalising decommodification indicates the thinking behind the concept. The method scores

> an array of variables that illustrate the ease with which an average person can opt out of the market: first, the prohibitiveness of conditions for eligibility, such as work experience, contributions, or means-tests; second, the strength of in-built disincentives (such as waiting days for cash benefits) and maximum duration of entitlements; and third, the degree to which benefits approximate normal expected earnings levels.
>
> (1990: 49)

Esping-Andersen builds his conclusions about types of welfare state from the scores outlined above for what he describes as the three most important social welfare programmes (pensions, sickness and unemployment cash benefits). This is an ambitious analysis but is deficient in two ways. First, it artificially separates off welfare programmes from the wider structure of the welfare state of which they are an integral part. The most important elements of this include fiscal and occupational welfare, the taxation systems, educational provision, arrangements for health service provision, housing arrangements and arrangements for water and transport services. The adequacy and generosity of the provisions Esping-Andersen analyses can only be assessed in the light of these other arrangements. The taxation system involves important elements of redistribution of income and a more complex analysis of these arrangements is required. In most countries systems of income taxation predate those of social security, and dependency between these two systems is considerable. The differential tax treatment of public and private insurance arrangements, including tax relief arrangements, often predated public programmes (Reddin 1970). Taxation modifies the impact of social security systems and the extent to which benefits are taxed affects the notion of insurance rights. Tax reliefs for occupational or private insurance represent state sponsorship and reduce the real market cost to households. These kinds of considerations cast doubt on the adequacy of Esping-Andersen's demarcation of social security arrangements. As Reddin (1970) has stated, any examination of social security systems relating to contribution incidence, value of benefits or wider issues of redistribution may be rendered meaningless unless such systems are examined in conjunction with other forms of benefits, services, labour markets and the tax system.

Definitions of the social security system also normally include provisions for health. Where health provision is provided by the market but access is dependent on insurance status, welfare benefits are designed in the light of this. In particular they are designed to enable people on benefit to maintain their insurance contributions. Where the health system is provided wholly outside the market, this affects taxation and benefit systems. Esping-Andersen attempts to analyse the extent of decommodification and the ease of opting out of the market without including health provision. These same points can be made in relation to education provision and whether charges are levied for education. Finally, in this context, Esping-Andersen fails to recognise rent control and rent subsidy schemes as integral parts of social welfare provision with direct implications for benefit levels. Interference with market rents is an alternative to generous benefits as a way of enabling the citizen to opt out of the market. It may be argued that social rented housing is itself an element in decommodification and loosens the pure commodity status of housing. Decommodification would seem to be high where standards of accommodation are at least as high as generally obtains in the market, where what is paid for housing is less than the market price, where there is no maximum duration of entitlement and where access and choice within the sector are not determined by income or indirect measures of income. Each of these elements would involve an extended debate and disagreements before the appropriate degree of decommodification could be proposed. Nevertheless, it is apparent that a decommodified housing sector forms as much a part of the apparatus of welfare regimes as benefit systems and in some countries affects a very large section of society.

These elements (health and housing most clearly) are particularly important in the British context where a National Health Service, free at the point of consumption, and low rents associated with rent control and subsidy have, in the past, been critical influences on benefit rates. These kinds of considerations cast doubt on the adequacy of Esping-Andersen's analysis as a guide to differences in patterns of urban change. Two other reservations can be expressed.

First, Esping-Andersen acknowledges that policies change but ultimately places insufficient emphasis on welfare restructuring and welfare regimes undergoing change. Second, and especially in relation to questions of segregation, the operation of the housing market needs some additional consideration. The existence of parts of the housing system which are decommodified – not just in terms of payments for use but also in terms of the negotiation of access – must be a key aspect of how far welfare states enable an average person to opt out of the market.

These considerations suggest that Esping-Andersen's analysis is seriously flawed and his operationalisation of concepts is insufficient for debates about urban change. An analysis of welfare state regimes based on a wider set of institutions would be more appropriate. However, typifying the whole of the welfare state is an over-ambitious aim involving drawing artificial boundaries within political economies. The retreat into ranking and counting also contradicts the terminology of rights and decommodification which does not easily translate into numbers. No countries wholly fit the emerging types and all have mixed systems of welfare reflecting a long history of partial reform. For the focus in this chapter, these elements and the failure to incorporate measures of active housing and planning policies which involve decommodification are serious problems. These issues are particularly relevant to how the British welfare state relates to those in the rest of Europe and North America. The view that Britain is best grouped with an Anglo-Saxon bloc in a liberal welfare state category is seriously in error. Socialised medicine and decommodified housing are in sharp contrast to these supposed nearest neighbours and represent greater decommodification than in the conservative/corporatist regimes and some social democratic regimes. Regimes are not so easily categorised. Nor are they static. Again, the British experience is one of active commodification or recommodification in housing, health, pensions and other areas. Finally, regional and local variations are considerable in some areas of provision – notably housing – and these territorial differences have to be addressed when our focus is at city level.

Segregation, exclusion and housing

British welfare state cities reflect a legacy of previous state policy interventions as well as the current structure of the welfare state. Both the built environment and where people live represent the outcome of individual decisions carried out in the context both of economic processes and of the welfare state. The direct state interventions in the housing market are key elements, along with health, income maintenance and other systems. These arrangements enable households to cope with crises which could otherwise force relocation and affect residential patterns. Welfare state arrangements, as well as family and reciprocal arrangements, mean that in Britain, as in the Netherlands (Musterd and Ostendorf 1994), where people live cannot be read off simply from income or household characteristics. Poor housing cannot be assumed to be associated with low income. Patterns of social segregation in Britain relate to welfare state arrangements which are not adequately typified as liberal or Anglo-Saxon. Leaving aside other elements, the social rented or decommodified housing sector has been more significant than other countries in this 'regime' type. Patterns of residence differ significantly from those in liberal or market systems without decommodified housing or with only small 'welfare' sectors.

This is an important backdrop to the present debate about the impact of growing social and income inequality on residential segregation and exclusion. Observation of labour market and economic changes which create greater inequality has created a tendency to assert that greater residential segregation follows. Thus, Winchester and White have argued (1988) that any group characterised by economic marginality (including the unemployed, impoverished elderly, students, single-parent families, ethnic minorities, refugees and handicapped people) will be constrained in their residential location and their economic status will be reflected in their occupation of the poorest sections of the housing stock. This equation gives too little attention to various coping strategies, welfare state, housing and other arrangements which affect how far and how fast economic change translates into change in residential patterns. Harloe, Marcuse and Smith (1992) offer a more complex and considered picture of patterns of housing change and emphasise the importance of restructuring housing provision and consumption as well as of economic changes. Their analysis lacks detailed spatial application but is a more appropriate starting point for debate.

In view of the importance of welfare state and housing arrangements for social segregation and inclusion, changes in these arrangements will be crucial for changes in segregation and for how economic change impacts on segregation. In British cities the rapid restructuring and commodification of the housing market has implications for segregation. In addition to the outcomes of economic changes or global processes there are also outcomes from the restructuring of the welfare state and local processes. Changes in the British housing market have changed the pattern of access and residential behaviour. The rapid decline of the private rented sector reduced the supply of easy access housing and altered the spatial association between lower-income people and poor-quality housing in the private rented sector. The state housing sector became the major accessible tenure for those who could not afford to buy housing (Murie 1983). At its peak in the late 1970s, council housing provided for almost one in three households and was still marked by considerable mix in terms of age, income, occupation and household type. It was a tenure of inclusion where lower-income households obtained better housing than was available to them in the market and many households able to buy chose to stay in council housing, which was generally modern, high-quality housing.

Over time different dwelling types and locations in the council sector developed different popularity, reputation, turnover and ease of access. Differentiation became significant. In the next phase of policy, with the sale of almost one in three council dwellings between 1979 and 1994, the dwellings sold were disproportionately the better dwellings. The transfer of this stock to home ownership has considerably altered that tenure and widened its social composition. At the same time, commodification of parts of the council housing sector narrowed its social base, with increasing proportions of households being in the lowest-income decile and declining proportions among households with incomes above the lowest two deciles (Murie 1993b). This pattern of change in council housing has been referred to as residualisation and commenced well before council house sales became significant. Since 1980, sales of council housing have developed in a context of wider commodification with rising rents and the removal of object subsidies. As rents have risen towards market rents, subsidised purchase has been a better option except for those on low incomes entitled to housing benefit. The proportion of tenants in receipt of housing benefit has grown and high marginal tax rates associated with withdrawal of benefits represent a poverty trap extended over a wider range of income. Council housing has increasingly become a tenure of exclusion – housing those who are excluded from other tenures and from the incomes and employment which enable inclusion in other tenures.

As the council sector has changed and declined in size, the private rented sector has stabilised in size and, latterly, housing associations have been encouraged. The rented sector has fragmented with different rights and rents and locations and qualities of properties. Commodification and the increase in market provision have complicated the characteristics associated with different parts of the housing system. Tenures are no longer so spatially concentrated and the expansion of home ownership has introduced more social mix into that tenure. Home ownership has more lower-income and elderly households and more who have only gained access to the tenure because of special incentives and policies. Decommodified or social rented housing, however, has become less mixed and caters more exclusively for the excluded.

The increasing proportion of skilled and other manual workers in home ownership has broken the particular pattern of socio-tenurial polarisation identified prior to 1981 (e.g. Hamnett 1984). This identified the increasing concentration of professional and managerial groups in owner-occupation and of the less skilled in council housing. By the mid-1980s British rented housing and council housing were increasingly associated with the non-working poor and home ownership had broadened its social base. This pattern is even more pronounced in the mid-1990s with rising homelessness, and problems in the home ownership sector (especially higher levels of mortgage failure and repossession) have represented new elements of exclusion from the dominant tenure and housing generally. The residualisation of council housing and similar patterns in housing associations and parts of the private rented sector, along with the different degrees of break-up of ownership within the council-built neighbourhoods, produces a more complex spatial pattern. As council housing has been sold and is resold, the social composition of council-built neighbourhoods changes – not always dramatically. In some cases, social class composition may change very little but in others there is a significant change (Forrest and Murie 1990, Williams and Twine 1990, Forrest, Murie and Gordon 1995). Privatisation does not equate with commodification and this may occur a long time later when market exchange takes place (Forrest and Murie 1995). Socio-tenurial polarisation with a focus on broad tenure categories and occupational class is no longer an adequate framework for analysing social or spatial outcomes of processes of change. It is no longer possible to read off spatial patterns from tenure patterns.

Social and housing change in Edinburgh

In the final section of this chapter the themes outlined above are pursued in relation to the City of Edinburgh, drawing on data from the Censuses of Population of 1981 and 1991. The focus is on the social and spatial patterns emerging in a period of change in the local economy and housing market.

Edinburgh is the historic capital of Scotland, initially built around a crag and tail formation providing a strong defensible site. The medieval city clustered closely around the castle rock and high ground around it. As a capital city it has had the full functions associated with the capital of a nation-state. Even following the Union of Scotland with England in 1707, the separate Scottish legal system and administration of government sustained governmental activities alongside the educational and financial traditions of the city. The administrative responsibilities of the Scottish Office established in 1885 have expanded since and have been an important element in consolidating the social structure of the city. Its economy has been based on legal, governmental, educational and financial services rather than on the relatively small manufacturing sector. As Hague describes it (1993: 7):

> It has always been a city of consumption and administration rather than a city of production. . . . In Victorian times its image was summarised in the epithet 'beer, beauty and bibles' signifying local service industries, tourism and education and administration. This economic structure begot a social structure that was disproportionately middle class, the most bourgeois town in Britain.

Lawyers, civil servants, academics and professionals working in financial services administering life insurance and pension funds were well established in the social structure of the city. During the 1980s these professions and services, and especially financial services, have grown considerably. At the same time, the relatively small manufacturing sector declined. Analysis of changes between 1981 and 1991 suggests a pattern of change broadly comparable with that noted for other cities. The employment structure has become even more skewed towards professionals working in highly desirable central area locations. A decline in manufacturing employment has occurred alongside rising unemployment (Hine and Wang 1994).

The population of Edinburgh within the city boundary declined by 1.5 per cent between 1981 and 1991 (from 425,256 to 418,914). Within the historic, highly desirable, residential areas of the old and new towns, population grew slightly in the period. The actual decline in population in council housing areas was greater than that of the city as a whole. Population growth occurred outside the areas of council housing mainly developed in the post-war period. The proportion of the city population in the 25–59 age group had increased from 43.2 per cent in 1981 to 47.5 per cent in 1991. Those aged over 60 accounted for 22 per cent in both periods and the proportion of younger households had declined. The city had become more one of people of working age. The proportion of residents aged 16 or over who were recorded as economically inactive rose from 127,976 in 1981 (37 per cent of all persons) to 132,320 in 1991 (38 per cent of all persons). The economically active population had declined and the unemployment rate in the city (unemployed as a percentage of all economically active persons) had risen from 7.7 per cent to 8.6 per cent with male unemployment rising more (9.7 per cent to 10.8 per cent) than female (5.2 per cent to 6.0 per cent).

The key indicators of economic change relate to this increase in unemployment and to changes in the respective roles of manufacturing industry and services. The percentage of residents aged 16 or over in the manufacturing sector declined from 16.9 per cent in 1981 to 11.3 per cent in 1991. In contrast, those employed in the service sector grew from 47.4 per cent in 1981 to 57.1 per cent in 1991. When data for the socio-economic group are considered, the number of manual workers has declined dramatically and that of junior non-manual workers has also declined. Other groups showed a numerical increase. The relative shares of different socio-economic groups show a considerable change over the decade.

This shift is similar to that which Hamnett (1994a), referring to London, has argued represents professionalisation rather than polarisation. Hamnett argues

Table 7.1 Socio-economic group of households (%)

	1981	*1991*
Employers and managers	14.5	18.9
Professional workers	8.4	11.5
Intermediate non-manual workers	12.5	16.8
Junior non-manual workers	16.7	15.7
Manual workers	25.2	19.3
Personal service and semi-skilled manual workers	13.9	11.5
Unskilled manual workers	6.6	4.7
Other	2.2	1.7
Number (10% sample)	13,247	10,927

that the absence of growth among the low-skilled occupations means that the picture cannot reasonably be presented as growth at two poles but is a pattern of professional growth only. Hamnett's argument is convincing if we restrict the consideration to these data. The misleading element in it is the exclusion from consideration of the unemployed and those who are not economically active. Professionalisation of those in employment has occurred alongside a polarisation between, at the one end, those who are not economically active, are unemployed, or are in low-paid employment and, at the other end, professionals and managers. This polarisation is a real one although it may be less resonant than reference to increased differentiation and the avoidance of artificial dualisms when there are more than two experiences of change. An additional important feature of the city has been the growth in the student population between 1981 and 1991 and its impact on the accessible parts of the private rented and owner-occupied markets.

These social changes have not occurred against an unchanging background. The housing market has changed significantly over the same period. Between 1981 and 1991 the census records an increase of 13 per cent in the number of households in Edinburgh. Changes in the tenure structure are much more dramatic (Tables 7.2 and 7.3). The number of owner-occupier households rose by 42 per cent and the number of council tenant households fell by 35 per cent. The proportion of households who were home owners rose between 1981 and 1991 from 53 per cent to 66 per cent and the proportion who were council tenants fell from 33 per cent to 19 per cent (Table 7.3). This change is the result of differential rates of new building between tenures, tenure transfers (especially under the Right to Buy), demolitions and changes in the numbers of vacant dwellings. Although the private rental sector had decreased slightly, the major shift to the owner-occupier sector came from transfers from local authority housing. In 1981 over 33 per cent of households (54,380) in Edinburgh lived in council or Scottish Special Housing Association (SSHA)-owned dwellings. By 1991 this had fallen to less than 20 per cent (36,950). The total share of the private rental sector has declined. But within this sector there was an increase in furnished accommodation and a decrease in unfurnished accommodation.

Changing council estates

In the remainder of this chapter the focus is on council estates and how commodification of housing in a period of increased unemployment is associated with spatial patterns of deprivation. The council tenure has seen a decline in its role, especially because of the Right to Buy, and this has involved a change in spatial patterns of tenure and in housing tenure on council-built estates. These changes have not been uniform and how this relates to patterns of social difference requires new analysis.

The initial step in the analysis in this chapter is to identify council estates. This could be done purely by reference to the 1981 census. However, there

Table 7.2 City of Edinburgh: households and housing tenure

	1981	*1991*
All households	164,692	185,664
Owner-occupied	86,830	123,352
Rented privately or with a job or business	18,345	18,002
Rented from HA	5,008	7,359
Rented from a council, new town or Scottish Homes (SSHA)	54,380	36,950

are problems with this in cities where there had been substantial council house sales prior to 1981. Equally important in this case was a wish to be able to identify estates which were built at different times and had other characteristics which are not included in census data but are relevant to patterns of change. The initial stage for Edinburgh was therefore to identify major council housing areas in the city on a base map. These council areas could be described in terms of when they were built, whether the estate included high-rise housing and what the extent of demand for the estate was. A link was then made to the census. This involved linking the boundaries of housing schemes with census areas. In the 1991 Census for Scotland, most 1981 Enumeration Districts (EDs) were subdivided into several Output Areas (OAs). In Edinburgh 1,492 EDs were subdivided into 3,592 OAs. For comparison between the two censuses, the analysis in this chapter used the 1981 EDs to build analysis areas which related to council estates. In determining final boundaries, reference was made to tenure in 1981 as well as to the original base map.

The key questions for this chapter are about change in the council housing areas referred to above. These areas have experienced a decline in the proportion of council tenants but this decline has not been uniform. They have also

Table 7.3 Percentage of households in different tenures in Edinburgh, 1981 and 1991

Tenure types	*Edinburgh City*		
	1981	*1991*	*% change*
Total	100.0	100.0	
Owner-occupied	52.8	66.4	13.6
Owned outright		21.9	
Buying		44.6	
Rented privately	9.6	8.6	–1.0
Furnished	4.1	5.7	1.6
Unfurnished	5.5	2.9	–2.6
Rented with a job or business	1.5	1.1	–0.3
Rented from a housing association	3.0	4.0	1.0
Rented from local authority	33.1	19.1	–14.0
Rented from Scottish Homes[a]	–	0.8	–

Note
a Scottish Homes (SSHA) was included with local authorities in 1981.

experienced other changes including those associated with general patterns of demographic and economic change over the period. Which areas have experienced most tenure change and how does this link with their social composition?

The initial part of the analysis refers to the 72 (71 for 1991 because of demolition) identified areas added together. In both 1981 and 1991 they accounted for over 80 per cent of all council housing in the city and the rate at which council housing has declined in these areas together is very similar to that for the city as a whole. As a result the data can be regarded as providing a robust picture of what has been happening in the city.

Of the council housing areas identified, 10 were built in the 1920s, 4 in the 1930s, 6 in the 1940s, 23 in the 1950s, 22 in the 1960s and 6 in the 1970s. They include 21 (30 per cent) areas of high-rise flats. The other areas are either of houses or tenement flats. In addition to these age and dwelling-type variables the analysis presented below refers to a popularity category based on an analysis carried out in 1978 by the city council Housing Department (City of Edinburgh 1979). From this, a variable was created for each of the council housing areas which divided them into four groups. This refers separately to tower blocks (21 areas) and divides the remainder into: popular areas (14), unpopular areas (15) and other areas (21). These data indicate that estates built in the 1920s were most likely to be popular (7 out of 10 areas) and that estates built in the 1970s were most likely to be unpopular areas (5 out of 6 areas).

In the selected housing estates the general pattern of change was similar to that for the city as a whole, with a decline in the public sector and an increase in owner-occupation. However, the shift was even more dramatic. The number of council and Scottish Homes tenants declined from 43,500 in 1981 to 29,300 in 1991. This has reduced the proportion of council tenants in these selected areas from 85.8 per cent to 56.7 per cent.

Table 7.4 indicates that areas identified in 1979 as unpopular had the highest proportions of children, unemployed persons, households without a car and vacant dwellings. Density of occupation was also highest. Only tower blocks had higher proportions of female lone-parent households. Tower blocks came second to the unpopular estates on all of the variables referred to above. The other two categories of estate scored very much lower on most of these variables but had higher proportions of economically inactive persons. In 1991 the areas which had been least popular in 1979 continued to have the highest child populations and the highest rates of unemployment and they had moved ahead of tower blocks in terms of female lone-parent households. However, in 1991 tower block areas had displaced the 1979 least popular areas in terms of lack of car ownership and density of occupation. The figures on vacant dwellings in 1991 reflect policy interventions to keep properties vacant pending refurbishment or demolition. The unpopular areas and tower blocks remained most like one another in 1991 and in terms of unemployment, economic inactivity and car ownership the gap between the two groups had widened. The unpopular areas and tower blocks

Table 7.4 A summary of major indicators for different areas of council housing

Indicators and year	*Tower blocks*	*Unpopular areas*	*Popular areas*	*Other areas*	*All areas*
Population aged 0–15 (%)					
1981	25.1	31.0	14.2	20.0	22.7
1991	21.3	28.4	13.9	18.8	20.6
Change	–3.8	–2.6	–0.3	–1.2	–2.1
Unemployment rate (%)					
1981	13.0	18.4	7.0	9.7	12.0
1991	22.4	23.2	9.6	13.3	17.3
Change	9.4	4.8	2.6	3.6	5.3
Persons aged 16 or over economically inactive (%)					
1981	34.0	34.5	39.5	36.3	35.9
1991	41.4	39.7	49.0	44.7	43.5
Change	7.4	5.2	9.5	8.4	7.6
Female lone-parent households (%)					
1981	8.4	6.6	1.6	3.1	5.1
1991	12.3	13.8	4.3	7.2	9.5
Change	3.9	7.2	2.7	4.1	4.4
Households without a car (%)					
1981	74.6	75.1	66.3	65.8	70.4
1991	76.2	72.4	62.4	62.1	68.5
Change	1.6	–2.7	–3.9	–3.7	–1.9
Vacant dwellings (%)					
1981	5.0	5.5	2.0	1.8	3.6
1991	17.3	16.0	2.6	6.4	10.9
Change	12.3	10.5	0.6	4.6	7.3
Households with 1.5 or more persons per room (%)					
1981	3.9	5.7	2.4	3.7	3.9
1991	1.1	0.9	0.4	0.7	0.8
Change	–2.8	–4.8	–2.0	–3.0	–3.1

were more distinctive in catering for younger families and households excluded from employment.

The broad implication of these data is of continuity. Tower blocks and the least popular estates were more deprived in both 1981 and 1991. This pattern is consistent with other evidence which indicates that the high rate of privatisation in the 1980s has not been associated with rapid social change. The key questions, however, remain about the pattern of change at the level of the individual estate. There is not room in this chapter to present this analysis in full and a summary only is offered here. The analysis carried out for individual estates has involved ranking estates in terms of *z* scores based on the variables indicative of deprivation (children under 15; unemployed persons; economically inactive persons; female lone parents; vacant dwellings; and overcrowding).

Rankings were prepared for 1981 and 1991 for the same areas. These rankings indicate that in 1981 the least deprived estates already had lower proportions of council housing than the most deprived estates. Very few of the least deprived estates were areas with tower blocks. In 1991 these areas continued to rank among the least deprived estates. The top 34 estates in 1981 were all in the top 42 in 1991. And it was in these areas that the rate of growth of home ownership was most marked between 1981 and 1991. In contrast, the most deprived estates in 1981 had much lower rates of growth of home ownership. They were more likely to be areas with tower blocks – but were not exclusively so. As they had started the 1980s with higher concentrations of council housing, they started the 1990s even more distinct from the other council neighbourhoods in this analysis.

The pattern which emerges is not one of total continuity. Some of the very worst estates in 1981 had been the object of major investment and other policy initiatives. The worst estate in 1981 had moved to position 36 in 1991, indicating the considerable social change associated with vacating and improving estates. The number of estates which had been transformed in this way was small and the estates which had replaced them at the foot of the rank order were those which were only just above them in 1981. There has been a commitment to council estates and to developing policy initiatives and the social and other characteristics of these areas do show changes as a result. The most dramatic changes in this respect are where policies involved displacement of population, as in West Pilton and in estates where demolitions have been significant. A full account of change would also refer to physical upgrading on estates and to changes in allocation and other policies affecting estates and designed to counteract the effects of earlier policies which had increased differentiation between estates. However, interventions have been, by necessity, selective and have taken place against a background of increasing social inequality and policies which have encouraged tenants who can afford to do so to purchase their dwelling.

The census analysis presented here enables the changes in housing through the 1980s to be analysed at estate level. The resulting picture is fully consistent with earlier analyses of the residualisation of council housing and patterns of social change associated with both residualisation and privatisation. In Edinburgh a major shift in the tenure structure has not affected all council estates to the same extent. There is a widening gap between the city as a whole and the council estates as a group. At the same time differences between council areas have become more marked. Two extreme cases emerge. The first is of more popular estates which are privatising rapidly. The 1991 census shows major tenure change but less dramatic social change. These are relatively stable and affluent areas and social change is likely to be slow and cumulative, with market exchanges accounting for an increasing proportion of residential movement and resulting in some gentrification with a comparatively affluent, younger population moving in. The second case relates to less popular areas and tower blocks which have established profiles more like residual estates. In these areas privatisation is occurring more slowly and patterns of social change relate to demand for council housing and processes of allocation and transfer within the sector. Households moving are likely to be disproportionately of unemployed lower-income and welfare-dependent households. In the short term privatisation has not resulted in major social changes but in the longer term the changes which are apparent are fully consistent with increasing differentiation between neighbourhoods which have been council estates. Census data identify where these changes are likely to occur and demonstrate the connections between dwelling type and other estate characteristics and the uneven processes of privatisation and residualisation.

These data raise important issues for analysis of patterns of social segregation and exclusion. Different council estates are on different trajectories and subject to different processes. Some areas are being consolidated as relatively affluent, popular and private. Other areas are increasingly excluded, with high and increasing concentrations of deprivation. The widening social inequality resulting from economic change is, and will continue to be, apparent in strengthening spatial divisions between council estates but these differences do not emerge purely as outcomes of economic change. They are fundamentally affected by the restructuring of the housing market and changes in housing policy and the wider welfare state.

Conclusions

Changes in the City of Edinburgh are compatible with the description of an increasingly unequal city. The widening social inequality which has characterised Britain since 1979 has not passed the city by. Economic change has resulted in an expansion in the numbers of those in professional and managerial jobs and a growth in unemployment and economic inactivity. Manual employment and employment in manufacturing have declined. Patterns of change have not been greatly affected by immigration and especially not of unskilled persons. The spatial outcomes of these processes cannot easily be understood without reference to the nature of the welfare state in Britain. The assumption that the British welfare state has involved limited decommodification is misleading and takes insufficient account of the wider structure of the welfare state, including a large decommodified housing sector. This wider welfare state has had a major impact on where people live and on how economic change affects patterns of residence. However, in the period since 1979, welfare state restructuring has been significant and more radical than elsewhere in Europe. If the post-war redistributive, integrative welfare state had already been eroded, the changes of the 1980s went further in the direction of creating a liberal welfare state. The legacy of earlier arrangements remains important, especially in where people live, but the continuing arrangements with more limited decommodification have a more limited redistributive effect. People entering the housing market for the first time are more likely to take different routes which relate to their economic position.

Privatisation and other policies have contributed to the dramatic expansion of home ownership. The process of this growth has involved discounted sales rather than market transactions and consequently has not initially been associated with changes of residence. The analysis of data for Edinburgh shows that those council areas which have seen most tenure change are those which were most affluent at the outset. Thus, those who have survived the restructuring of the economy have disproportionately transferred their dwelling to the market sector and by doing this reduced the extent of decommodified provision. What is left in council housing reflects the distribution of those who have not fared well economically as well as aspects of age and housing features. The growing population experiencing unemployment or outside the labour market are dependent on a smaller, more fragmented rented market which has a different spatial pattern from that of the past. The spatial outcomes of global processes, and the key factors which determine where people differentially affected by economic change live, crucially concern the operation of the welfare state and the housing market and their restructuring. In the British context, where such restructuring has been substantial and significant, its importance in changing patterns in cities may be greater than either where liberal welfare regimes have always had a limited impact on the extent to which the individual can opt out of the market, or than where redistributive welfare states have remained relatively unscathed. The emerging pattern is one where those who are not homeless or excluded from the labour market are more likely to obtain housing through the market. In some cases this involves residence in council-built areas which are absorbed into or included within new social boundaries. In contrast, other council and social rented housing has an increasingly distinctive social role relating to position in the market. These areas are becoming or have become areas of exclusion. Living in such areas further disadvantages people and a downward spiral exacerbates the spatial division within cities. Those who have not shared in the opportunities resulting from economic change are more restricted in choices and the residential structure of the city is becoming one in which economic position, housing tenure and residential location are more strongly linked. Understanding this requires consideration of housing and welfare state systems as well as global economic pressures.

Acknowledgements

I am grateful for help and advice from Dr Lou Rosenberg and Dr Ya Ping Wang in carrying out the analysis for this chapter.

8

THE GEOGRAPHY OF DEPRIVATION IN BRUSSELS AND LOCAL DEVELOPMENT STRATEGIES

Christian Kesteloot

Introduction

As has been observed in earlier chapters, many theoretical and empirical studies of world cities have shown the existence of new mechanisms involved in the social production of poverty. The main argument is that the new division of labour, linked to the globalisation of the economy, generates simultaneously an increase in highly skilled and well-paid jobs and in unskilled, unstable jobs that are insufficiently remunerated to provide a satisfactory standard of living. This dualisation or social polarisation thesis, first expounded by Friedmann and Wolff (1982) and developed by Sassen (1991), has been questioned as over-simplifying the observed trends (e.g. Marcuse 1989, Hamnett 1994a and in this book) or underestimating national differences resulting from the actions of nation-states (Silver 1993). Nevertheless, it remains an empirical fact that in many western cities recent economic restructuring has enhanced social inequalities, even if they are not related to the globalisation processes.

However, explaining these processes remains a complex task. At the empirical level, research has to cover a very broad range of topics, such as changes in participation in the labour market; changes in the distribution of incomes, education, housing, health and culture; and variations in state intervention in all these fields. At the theoretical level, there is a growing consensus about three distinct spheres in which the sources of polarisation originate, namely transformations in the division of labour already mentioned; the restructurations of nation-states, and particularly the slow dismantling of the welfare state; and finally the second demographic transition, which results in the appearance of new household forms and the parallel increase of single persons and social isolation (Mingione 1996). These societal changes result in differentiating the access of households to the monetary and non-monetary resources necessary for their existence. Relative poverty, and often absolute poverty, is increased.

However, there is also another empirical entry to discussions of polarisation, namely socio-spatial polarisation. This process implies the existence and growth of spatial segregation between the rich and the poor. Spatial structuring becomes an active element in the process of social polarisation if rich and poor concentrate respectively in rich and poor environments in terms of the resources of collective consumption, housing, mobility and access to jobs. Of course, an increase in spatial segregation can theoretically occur without social polarisation. But in societies with a low level of public intervention in the socio-spatial field (concerning the provision of adequate housing, public transportation, education and sports facilities, cultural infrastructure, and so on),[1] spatial polarisation would generate social polarisation through the field of collective consumption.

In this chapter I shall focus on the manifestations of polarisation and segregation in the Brussels urban region, resulting in the existence of deprived neighbourhoods. I shall further try to show how important even small differences between these neighbourhoods can be for the creation of adequate territorial policies, using a framework for analysis derived from the concept of mode of economic integration introduced by Polanyi (1944) and later developed by Mingione (1991). In the first section, changes in the Belgian welfare state and their contribution to rising poverty are examined. Socio-spatial polarisation in Brussels is then analysed at three different spatial scales. The third section is devoted to the actual segregation processes leading both to the existence of deprived neighbourhoods in the city and to the nature of the differences between them. Finally, the chapter considers appropriate local development strategies based on spheres of economic integration.

Poverty and the changing welfare state in Belgium

Since the mid-1970s Belgium has been a fully developed welfare state, at least in the field of incomes. In 1974 the law on minimum incomes ensured subsistence to every Belgian adult. This law was seen as the final step in the national social security system. Whilst this secured minimum income was subjected to a series of conditions (such as willingness to work and the exhaustion of other social security rights), the law was complemented in 1976 with a new measure dealing with social assistance, proclaiming in its first article that every person has the right to social assistance in order to reach the conditions necessary for living in human dignity. The 1974 benefit was restricted to Belgians, but the 1976 addition accords the right to a minimal income to all residents of the country.

Unlike the French system of *revenu minimum d'insertion*, which was created as a tool against polarisation in the late 1980s and is associated with an individual contract of social integration, both the Belgian laws should be seen as the final elements of a welfare state conceptualised at the end of the Second World War and fully developed during the Golden Sixties. However, the economic crisis and the following period of economic growth without job creation have forced the Belgian system gradually to be bent towards the French model by incorporating active assistance aiming at socio-economic reinsertion. This was done as a result of the pressure produced by growing numbers of persons falling through the social security net with high and persistent unemployment levels, and because of the financial crisis of the state, of the arrival of asylum seekers from the second half of the 1980s and finally of the lowering of the age of majority from 21 to 18 in 1990.[2] The most significant steps in the process of welfare adjustment occurred in 1984 when the municipal welfare agencies (OCMW-CPAS), responsible for the distribution of the minimum income and social assistance, were compelled to reclaim financial support from relatives of the recipients, and in 1993 when individual contracts of social integration were made compulsory for young adults (18–25 years of age) applying for the minimum income benefit.

Recent evidence shows that social security benefits in Belgium can still prevent households falling into poverty, although the percentage of GDP allocated to these benefits has fallen from 19.9 per cent in 1985 to 16.9 per cent in 1992, whereas the numbers of recipients in all categories have increased significantly. The reason for this lies in the composition of households in which a social security income is increasingly combined with other sources of income.[3] Nevertheless, single persons, one-income households and one-parent families form one group, and young adults another group who are the main victims of these changes in welfare provision. The first group suffer because a drop in their income cannot be compensated by income from another source, whilst the second group suffer because of the lowering of the age of majority and the increasingly difficult step between education and work. Thus the figures for Belgium show a surprising decline in the proportion of the population living below the poverty line[4] (from 22 per cent of households in 1985 to 18 per cent in 1992), and the number of poor appears to be constant (using the official minimum income as the criterion they account for 3 per cent of households; using the European Union norms the figure is 6 per cent), but the number of minimum income recipients is increasing sharply (from 44,000 in 1986 to 75,000 in 1996) (see Vranken, Geldof and Van Menxel 1996).[5]

Social and spatial polarisation in Brussels

In Brussels, the same trends are present but even more extreme. In 1993, the proportion of households below the poverty line was 26 per cent, the proportion with below the official minimum income was 5.5 per cent and the number of minimum income recipients nearly doubled from around 6,200 in the late 1980s to 11,913 in 1995 (or 1.2 per cent of the Brussels population).[6] Moreover, social aid equivalent to the minimal income was granted to 8,131 non-Belgian residents. Including the guaranteed income scheme for the elderly, 5.6 per cent of the Brussels population depend directly on one of these residual welfare schemes for their income (De Keersmaecker 1997).

These figures show that the concentration of poverty is higher in Brussels than elsewhere in the country and that poverty is increasing. This provides only partial evidence for social polarisation, unless it is understood as a widening gap between rich and poor.[7] Indeed, the figures concern only the people and households who actually benefit from the residual schemes of the social security system. But many others, such as the long-term unemployed, disabled persons, or even people trying to become self-sufficient outside the market and state benefits, are experiencing deprivation without appearing in these statistics. Moreover, in terms of the dynamics of those close to the poverty line, a large group of urban youngsters, especially from migrant origins, is facing a gloomy future with respect to employment and income security (Kesteloot 1995a and b).

It could be that the changes just described result from growing poverty rather than from polarisation. Indeed, the incomes of the rich could be changing in the same direction. However, income figures do sustain the social polarisation thesis (Figure 8.1). Growth indices of the average taxable income of all tax-payers for the three lowest and three highest deciles clearly disclose polarisation tendencies from the 1980s. While in the 1970s lower incomes grew faster than the highest incomes, at least in relative terms, the tendency was reversed during the early 1980s, with a strong polarising upsurge between 1984 and 1988 and a minor one in the 1990s.[8] Moreover, a slight increase in the concentration of low and high incomes over the last years can be demonstrated.

Such social polarisation does not directly support the hypothesis of a dual city or a global city, since little is said about the macro-social processes leading

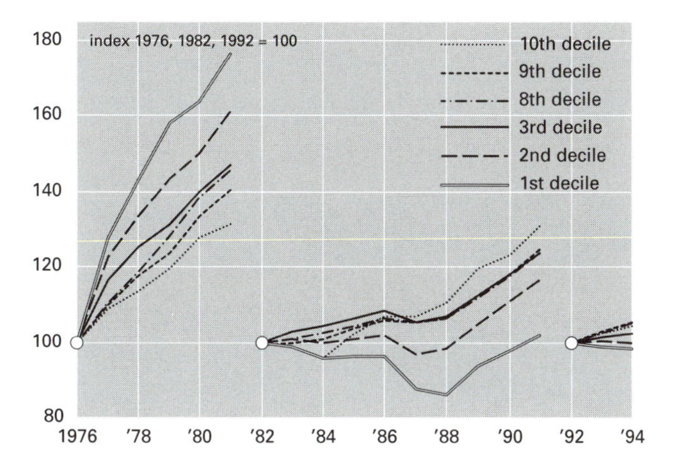

Figure 8.1 Changes in taxable income per taxpayer, by decile, Brussels Capital Region, 1976–94

to this widening gap between rich and poor. However, these changes have a profound impact on the dynamics of the social, economic and political structure of Brussels, because they are spatially translated into deepening segregation between the social groups. This spatial translation of social polarisation is manifest in the Brussels urban region[9] at three levels, which involve different socio-spatial processes (Kesteloot 1994).

The first process is suburbanisation, which created a deep spatial segregation along socio-economic, demographic and ethnic lines between the inner city and the periphery within the urban region. Thus, the vast majority of the poor neighbourhoods are located in the nineteenth-century belt of the city and predominantly on the northern and western side of it. This area has always been a working-class zone ever since the origins of industrialisation, while the eastern sector with hills and woods is more attractive for the middle and upper classes (Figure 8.2). Both the poorer elderly and the labour immigrants from Italy, Spain, Morocco and Turkey with their large young families reside in the inner city. They are temporarily joined by young starters on the urban housing market. In contrast, the highest income areas are mainly outside the Brussels Capital Region, particularly in the south-east and the north-west – again more attractive areas which are perpendicular to the NNE–SSW industrial axis along the canal linking Brussels with Antwerp and Charleroi. Typically the inhabitants of these two high-status sectors are Belgian middle- and upper-class families with children, joined by affluent expatriates with families, whose presence in Brussels is related to the international functions of the city (Kesteloot and Van der Haegen 1997).

The second process is the consolidation of the characteristics of ethnicity and poverty of the nineteenth-century inner-city neighbourhoods, through housing market mechanisms and the impact of economic changes. On Figure 8.2 this is clearly illustrated by the lowest-income area, sometimes called the poor crescent or the poor banana of Brussels. It coincides with the areas of concentration of Moroccan and Turkish populations (respectively 77,000 and 21,000 inhabitants, the former occupying the whole area, the latter being concentrated in the northern parts of it). The first wave of labour immigrants in the early 1960s came from Italy, Spain and to a lesser extent from Greece. They settled in the same areas but concentrated on the southern side of this poor crescent, since they arrived by train in the South station. Meanwhile, suburbanisation of the relatively successful second generation and return migration of the elderly have slowly eroded the South European presence in the area and opened the way for Moroccan dominance in the urban landscape.

The last process is not visible on Figure 8.2, since it concerns a few individual neighbourhoods displaying the same characteristics as other nineteenth-century districts, but which have experienced a downward spiral of environmental and social decay, bringing them to the verge of being 'no-go areas' in the city. These neighbourhoods (amounting to about 15 statistical sectors) systematically display very high scores on deprivation criteria (Kesteloot, Mistiaen and

Figure 8.2 Average taxable income per person, Brussels urban region, 1993

Decroly 1997), but they are chiefly distinguished by having the highest proportions of children and teenagers. These teenagers are of immigrant origins and the socialisation processes affecting them are hampered by economic changes, ethnic discrimination, cultural distress and political neglect (since non-Belgians have no political rights).[10]

Thus, in the Belgian urban context poverty is primarily an inner-city phenomenon.[11] In contrast to Southern European, or even Dutch and British cities, the Belgian urban environment is characterised by a sharp socio-economic division between the poor nineteenth-century belt and the wealthier suburban regions. Poverty in the urban periphery is almost absent (with the exception of a few, small and not very peripheral social housing estates – see below). Moreover, the concentration of poverty in the inner city accompanies the concentration of immigrants. Their share in the figures concerning poverty and as beneficiaries of residual social security schemes is negligible in most cases, but generally speaking they are, as a substitute for the Belgian urban working class, the main victims of unemployment resulting from economic stagnation, deindustrialisation, the relocation of economic activities and increases in productivity. Thus, unemployment among the active population is 33 and 30 per cent for Moroccans and Turks respectively, against 8 per cent for Belgians (1991 census figures). While the concentration of immigrants in the poor areas of the cities is of paramount importance in the problem of social exclusion and spatial segregation (see Figure 8.3, which illustrates the correspondence of the spatial concentration of guestworkers and their descendants with the poor areas

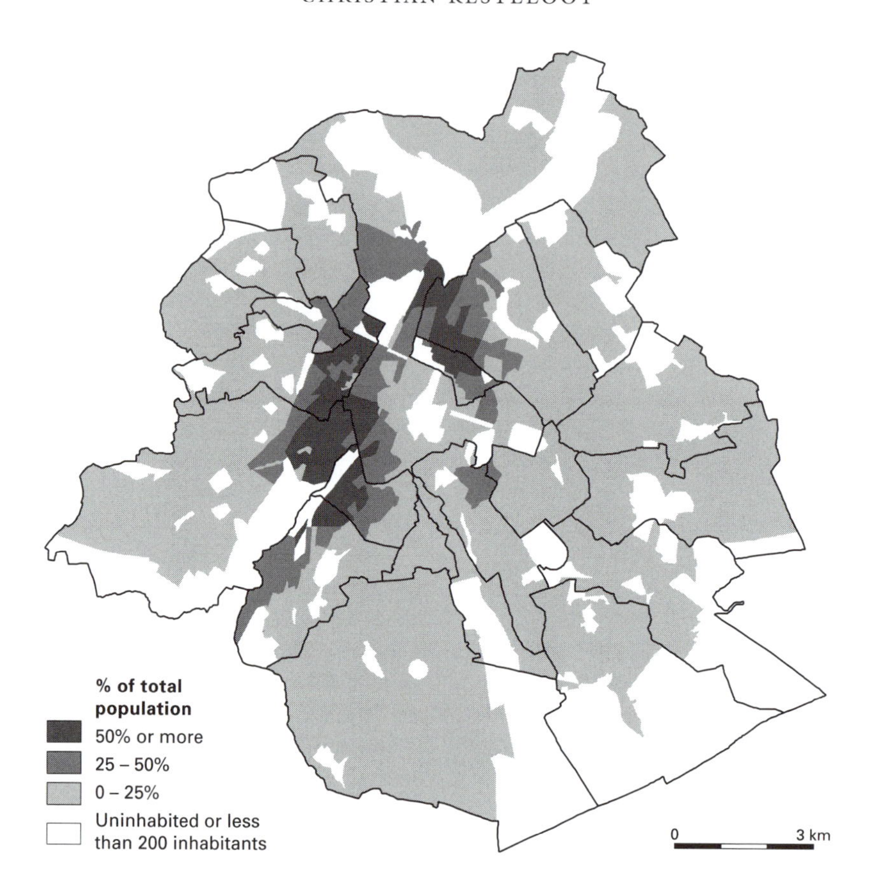

Figure 8.3 Distribution of guestworkers' nationalities, Brussels, 1991

on Figure 8.2), they are under-represented in the poverty figures because of discrimination on the one hand (strongly induced by the absence of political rights) and because of their relatively stronger resistance to poverty through their own social networks (Kesteloot *et al.* 1997).

This spatial pattern is evidently created by the extent of mass suburbanisation. Indeed, suburbanisation was a strong social and spatial process in the 1960s and the early 1970s, linked to economic growth and the general upward social mobility of the Belgian population in terms of education and incomes. Job security and steady wage improvements encouraged the middle class into home-ownership, and the absence of any restrictive planning regulation, together with state incentives for owner-occupation, channelled this movement into private house-building on cheap land in the urban periphery.

Unskilled labour migrants, called 'guestworkers' at that time, arrived from the second half of the 1960s to fill in both the socio-economic and spatial gaps left by the suburbanising population. As an inexpensive labour force, they took over jobs in the construction and transportation industry and in low-paid, labour-intensive services such as hotel, catering and cleaning services. These immigrants occupy what can be called the residual rental housing sector – the private rental sector offering cheap housing in old buildings (about 30 per cent of the Brussels housing market). This sector is dominant in the nineteenth-century inner-city neighbourhoods, abandoned by the majority of the Belgian population (Figure 8.4).[12]

These guestworker neighbourhoods have been consolidated as a result of the economic crisis which prevented both any upward social mobility and the

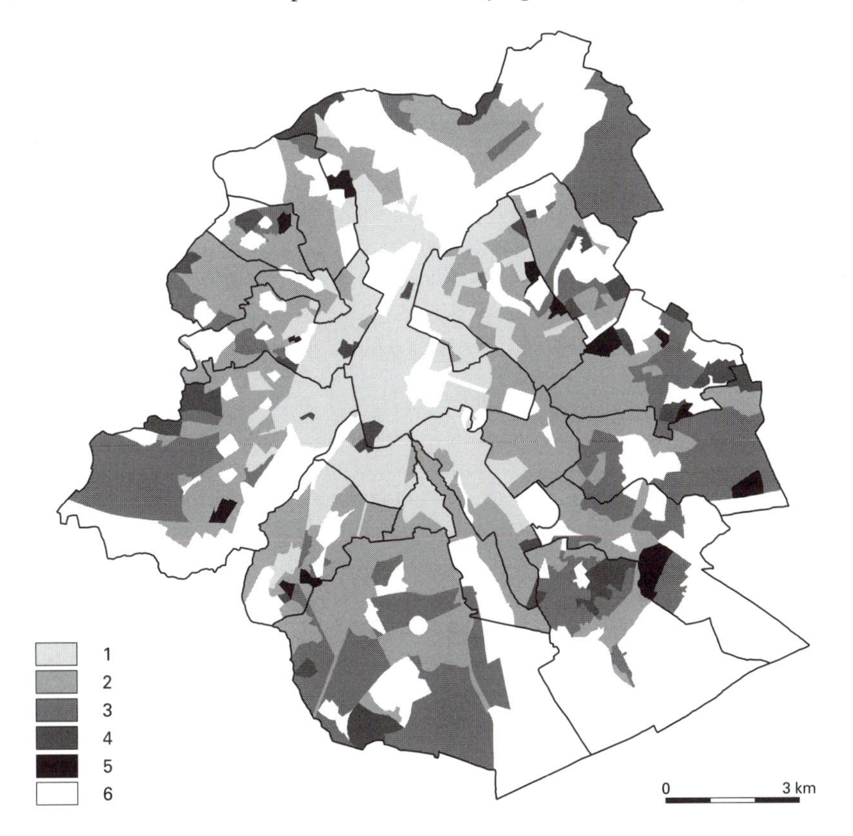

Figure 8.4 The spatial structure of the housing market in Brussels, 1981

1 Dominance of residual private rental sector and secondary owner-occupied sector (75 per cent rented dwellings or more)

2 Dominance of private rental sector (between 50 and 75 per cent rented dwellings)

3 Dominance of owner-occupied sector and better quality rental sector (50 per cent or more owner-occupied dwellings)

4 Social housing estates (25 per cent owner-occupied dwellings or more)

5 Social housing estates (less than 25 per cent owner-occupied dwellings)

6 Less than 200 inhabitants

gradual deconcentration of these groups from their original districts. Indeed, any change in residential location would entail higher housing costs, loss of access to the ethnic infrastructure and sometimes the abandonment of local labour opportunities. Moreover, many Belgian households were also restrained from further suburbanisation, which barred any possibility of a filtering-down process. Indeed, the crisis has kept many middle-class households – who otherwise would have become suburbanised – in the city. Some of these people remain in their urban neighbourhoods by buying and renovating relatively cheap housing. A similar outcome in the inner-city housing market results from the increase of single people and couples without children, which more than compensates for the loss of population in terms of the number of households.

Clearly, these processes have deepened the socio-spatial contrasts between centre and periphery in the urban region. While this urban duality has to be attributed to the economic crisis, the temporary recovery of the mid-1980s did not produce a reverse trend, but deepened the socio-spatial contradiction as a result of urban restructuring (Kesteloot 1995a, Kesteloot and Van der Haegen 1997).

Segregation processes and the geography of deprivation in Brussels

The three processes have quite different time and spatial scales. Suburbanisation affects the whole urban region and covers the whole post-war period. The consolidation of the ethnic neighbourhoods was triggered off by the economic crisis and is necessarily confined to the inner city, whilst the decline of some neighbourhoods into deeper deprivation is related to recent urban restructuring and highly localised factors. Nevertheless, taken together, these processes are the major causes of the concentration of poverty in certain areas of the city.

However, they do not constitute a universal explanation. As implied by the existence of some neighbourhoods on the verge of being no-go areas, the poor inner city shows considerable differentiation. This is important since it points to variations in the causes of socio-spatial polarisation and also suggests the need for differentiated policies. Segregation processes, providing the link between social deprivation and the spatial concentration of poverty, are the starting point for the analysis of these differences. The link is established through the mechanisms of the labour and housing markets. Three structural mechanisms of segregation can be identified, translating social deprivation into its spatial counterpart and thus leading to neighbourhood differentiation.

The most obvious mechanism is the spatial segregation of the poor. In the absence of strong state intervention in the housing field, the structure and functioning of the housing market is responsible for the spatial distribution of social groups. In Brussels, three sectors of the housing market have an impact on the spatial segregation of poverty, leaving on one side the limited sector of rented furnished rooms (Figure 8.4).

The most important sector, as mentioned earlier, is the residual private rental sector, which largely results from suburbanisation. Given the very limited role of social housing on the Brussels housing market (accounting for only 8 per cent of the housing stock), this residual sector offers the main accommodation possibilities for both poor households and starters in the housing market.

A second important sector is composed of residual owner-occupied property. As a result both of speculation (which doubled the cost of housing in Brussels between 1988 and 1992), and of the growing general feeling amongst immigrants that return migration makes no sense, more and more immigrants decided to buy their residences. In 1991, 37 per cent of Turkish and 30 per cent of Moroccan households were owner-occupiers, against 13 and 10 per cent respectively in 1981. This shift to owner-occupation did not entail a spatial shift. Most purchasers bought houses in their original neighbourhoods, sometimes even their own rented house, thus shifting dwellings from the residual rental sector to a residual owner-occupied sector. Inevitably this shift is a main element in consolidating the ethnic neighbourhoods of the city. It can even be seen as a strategy by which immigrants achieve territorial security, remaining in their original neighbourhoods and maintaining their access to the ethnic infrastructure, to social networks and to inner-city locations, which help them to survive under relatively poor economic conditions (Kesteloot, De Decker and Manço 1997). Taking into account the very high transaction costs in the Belgian housing market (amounting to up to 25 per cent of the dwelling price), this residual owner-occupied sector not only generates but also perpetuates spatial segregation.

Finally, the small social rental sector creates clear concentrations of some categories of the poor because of the larger scale of the estates. Older estates may house the poor elderly, alongside 'problematic' families, including migrant families – although this depends on the allocation policy adopted by individual social housing companies. In Brussels, the managers of a few of these estates accepted immigrants because of unfilled vacancies resulting from the architectural features of the estates: high-rise, poorly insulated apartment blocks with flats for large families, built just before the oil crisis. Typically, they are located in the inner suburbs of the urban region, within the territory of the Brussels Capital Region, although some of them are inner-city projects erected during slum clearance schemes.

A second segregation mechanism is social polarisation in circumstances where it shows spatially differentiated effects. At root, social polarisation is caused by changes in the labour market, including both the disappearance of certain activities and the transformation of others due to technological restructuring. It is reinforced by changes in state redistribution schemes (such as income progressive taxes, social security schemes and selective investments in collective infrastructure and services). The role of social polarisation in spatial deprivation is clearest where areas start to decay because they largely depend on one employer or economic activity which faces crisis or closure. Obviously this mechanism is more important on the regional than on the local scale, but because of its interaction with housing market processes, it can indeed translate into local differentiation. In Belgium, the problems of deprivation in the Walloon and the Limburg mining areas are the best known examples. This implies that larger cities, with their diversified economies, show more resistance to this mechanism. Since the onset of economic crisis, however, deindustrialisation and relocation play a similar role in intensifying the effects of poverty in the housing market. The more important the industrial activity in the employment structure before the crisis, the greater these effects will be. In the Brussels Capital Region, where service activities strongly dominate (with 90 per cent of total employment), technological restructuring and the effects of internationalisation are the most important factors creating social polarisation. They lead to the disappearance of low-skilled but stable industrial employment, the increase of low-skilled but highly insecure jobs in urban services, and a general tendency towards the suburbanisation of employment, leading to an employment mismatch (for more details see Kesteloot 1995a).

Third and finally, there are a number of concentration mechanisms through which social and spatial deprivation coincide. The clearest case occurs when social deprivation causes spatial deprivation, which then reinforces the former process through the interaction of negative effects in both the labour and housing markets. The spatial segregation of a deprived area then has an effect on the chances of securing a good position in the labour market, while social polarisation can cause selective emigration, until only the weaker groups continue to stay in such neighbourhoods. This interaction locks poverty into certain neighbourhoods, so that the problem of deprivation becomes much greater than the sum of the individual characteristics of deprivation shown by the inhabitants. Signs of this include: the development of schools in which migrants are concentrated, the absence of suitable social services and amenities, and the fact that local authorities neglect the area and allow it to decay. Finally, the absence of positive models and references in the neighbourhood hampers the socialisation process of young people (De Decker 1994, Mistiaen 1994).

Spatial structuring from the past continues to influence contemporary processes of socio-spatial polarisation, just as processes on one spatial scale interact with those on other scales. These interactions go further than those described here, since even the history of the creation of working-class areas over a hundred years ago still plays a role in contemporary indicators of deprivation. It must also be noted that the three segregation mechanisms influencing the spatial concentration of deprivation do not operate identically everywhere, with the result that the outcomes involve a very complicated typology of deprived areas in the city.[13] Small variations in the combination of past and present processes, or in the strength with which these processes affect particular areas, may cause important differences in deprivation, and more crucially determine the opportunities for development strategies.

International comparisons focus too much on the common characteristics of deprived areas (see, e.g., Jacquier 1994), whereas it should be evident that the social and spatial context of deprivation in the nineteenth-century belts of Brussels is very different from that of the British 'peripheral estates' and the French *grands ensembles* of the 1960s and 1970s. Neighbourhood development as an answer to marginalisation processes should keep in mind that deprived areas cannot all be dealt with in similar ways. Modalities for the reproduction of deprivation and the chances for emancipation may vary because of different causes of deprivation and as a result of different social and spatial contexts. Even in the Brussels context, differences remain important enough to cause variations in consequences, thus requiring neighbourhood development strategies that are appropriate for different circumstances.

Appropriate local development strategies: a framework for analysis

Market exchange, redistribution and reciprocity

Deprivation can be considered as a problem of insufficient access to the socio-economic resources necessary for a decent living and the reproduction of the household. Most of these resources are not produced directly by households, but by producers engaged in the economic system. Thus, access to these resources is not direct, but dependent on the integration of households within this economic system. This problem of access to resources has been theorised by Polanyi and is encapsulated in his concept of modes of socio-economic integration (1944). Based on his economic-anthropological work, the concept distinguishes three fundamental ways to obtain resources produced in common: reciprocity, redistribution and market exchange (see also Harvey 1973 and Mingione 1991).

In the western world, access to resources is dominated by market exchange. In a simplified form, this means that individuals and households must develop a social utility, i.e. they must produce goods or services needed by others. This gives them an income which allows them to buy the goods and services they need but which they cannot produce for themselves. Most households put their labour on the market. Their wage is the price they get when they succeed in selling it. Others are self-employed and sell goods and services. An important characteristic of market exchange as a factor of economic integration is the autonomy of the individual actors. They decide by themselves what they will bring to the market, and the law of supply and demand will indicate whether they made a good decision or not. Bad decisions are sanctioned by loss of income and reduced access to the resources produced by others. As a result, the market generates stratification and unequal access to resources, based on strong or weak positions in the market.

These inequalities, especially those related to households which can only develop a social utility by making their own labour available to others, are inherent in market exchange and can be socially destructive, since households without utility (whose labour is not needed) have no access to resources. This produces a structural reason why these inequalities are partially compensated by state redistribution.

From the households' point of view, redistribution means that everybody contributes to a common stock of resources, and that these are then redistributed according to a set of rules. Hence, redistribution implies a central collecting agency and a hierarchic organisation. The slow but steady development of the welfare state from the first social legislation, public education and the beginnings of public transport at the end of the nineteenth century, through social housing during the inter-war period, to the development of a comprehensive social security system after the Second World War, created massive and positive (from rich to poor) redistribution in Belgium, mainly fuelled by taxes and social security contributions. But participation in redistribution is largely determined by participation in the labour market. Nevertheless, in contrast to other means of economic integration, redistribution is the only one which can guarantee, without exceptions, access to basic resources for everybody, since the state has the power to decide that this should be so. This has been partially the case in Belgium, with the creation of the minimum income system and the right to social aid, considered in the first section of this chapter.

Finally, reciprocity helps people to obtain resources through mutual exchange. It implies that each of the participants has the capacity to produce some resources as well as the existence of a social network with symmetric links between members. Everything (goods and services) which is brought into the system by someone is exchanged with the other members, generally under the form of different goods and services, and very often not at the same moment. These features of the exchange process involve mutual trust between the members of an exchange network and lasting ties between each member and the network. This type of relationship to a network is termed 'affiliation'.[14] Hence, the most common networks are the household, the extended family and sometimes networks among neighbours or ethnic communities.

These three forms of economic integration have each been dominant in different periods of history, but this brief description illustrates that they are also all present in contemporary capitalist society. Indeed, they demarcate the different types of welfare state. Thus, redistribution predominates in the social-democrat model, the liberal welfare state emphasises market exchange and the conservative type is based on reciprocity (Esping-Andersen 1990). Current economic developments are mostly interpreted as cutting back the welfare state or, putting it in other terms, dismantling the redistributive role of the state in favour of the enhancement of market exchange. Reciprocity increases amongst those groups which do not hold a strong position in the market but which have good social networks. This type of society has been labelled a 'fragmented society', which does not mean a broken or disorganised society, but one in which the social organisation of household reproduction is increasingly complex (Mingione 1991).

Each mode of economic integration requires a different set of socio-spatial conditions. Therefore, development strategies and policies against poverty at the local level, which necessarily involve one or more of the modes of economic integration, will be favoured or hindered by the features of each neighbourhood involved. In general, it can be said that market exchange is applicable to the inhabitants of affluent peripheral neighbourhoods. In contrast, only a relatively small proportion of households in deprived areas can make a living out of incomes earned on the formal labour market. In most cases, this income is enlarged with benefits obtained by other members of the household, and with cost-cutting mutual services exchanged with other inhabitants or related families in the neighbourhood. But there may be areas where everybody is poor and can survive only through state redistribution or the use of social networks. Consequently, all three modes of economic integration are present, although the weighting of each in the mix varies according to the specific neighbourhood. This is even true at the household level. Hence, it is more accurate to speak about three spheres of economic integration, each containing some of the activities of the households, and aimed at securing their means of existence.

The problems and potentials of deprived neighbourhoods: survival strategies and territorial policies against poverty

In the previous subsection the concept of the three modes or spheres of economic integration appears as a very powerful tool for the understanding of the social and spatial dimensions of contemporary urban poverty. The concept is examined here in relation to the causes, effects and policies concerning poverty within each sphere and the differentiation of deprived areas in the city (Figure 8.5).

In deprived neighbourhoods, the socio-spatial conditions for the operation of modes of economic integration can be described in a negative way in terms of weaknesses hindering successful access to the goods and services necessary for making a decent living. Some of these weaknesses concern residential space as such and its embeddedness in the city; others concern its population. Empirical research on deprived neighbourhoods in Flanders and Brussels has revealed that these weaknesses are manifest in four domains (Kesteloot *et al.* 1996): social isolation (indicated by a high rate of single persons), poor access to goods and services redistributed by the state (especially for immigrants, in the absence of political rights), difficult access to the labour market (indicated by a high unemployment rate, but also through poor education levels) and finally bad housing conditions (which can affect both tenants and owner-occupiers in the neighbourhoods concerned). As expected, these four domains can be reduced to the three spheres of economic integration. Conversely, most of the positive elements in these neighbourhoods can be subsumed within opportunities for survival strategies, which can again be situated in each sphere (Kesteloot *et al.* 1997).

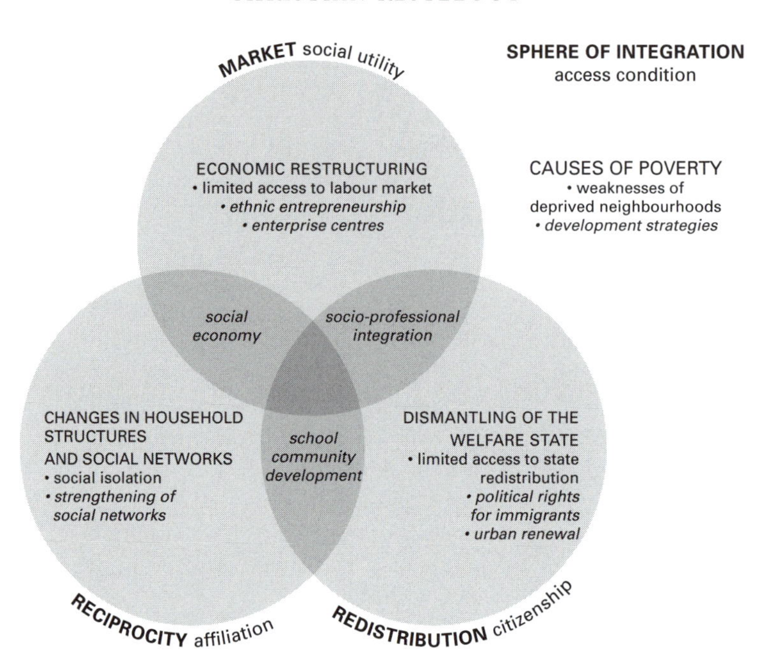

Figure 8.5 Spheres of economic integration, poverty and local development strategies

Social isolation, related to demographic changes and the rapid transformation of household structures but also to the weakening of local community life, mainly has effects on reciprocity. Single persons and one-parent families often have a lower capacity for the production of the means of existence within their own domestic sphere, and usually do not have a strong internal reciprocity network within their household and family. Moreover, most single persons are elderly or young persons, involving generations who are traditionally recipients of inter-generation reciprocity (the young receiving from their parents and giving later during their active life; the elderly having given during their active life and expecting to receive back from younger generations). With increasingly uncertain access to employment, the fact of having only one income source weakens these households. Since state redistribution is also decreasing, they are pushed into reciprocity strategies. But, paradoxically, access to resources through reciprocity is much harder for those who depend most on such access (see Pahl 1984 for this argument concerning survival strategies). However, well-structured ethnic communities which are certainly handicapped in the spheres of market exchange (through discrimination) and redistribution (see below) can compensate for this handicap through the sphere of reciprocity.

This is illustrated in Brussels by the Turkish community in the northern part of the poor crescent of the city (see Mistiaen, Meert and Kesteloot 1995 and Meert, Mistiaen and Kesteloot 1997). The Turks are spatially very concentrated (with a segregation index of 72 per cent) in a zone which nevertheless remains multicultural (consisting of an average 23 per cent Turks, the same proportion of Moroccans and 35 per cent Belgians). The fact that Turks tend to stick together also implies that their socio-economic differentiation is growing, allowing for solidarity between the rich and the poor. Finally, chain migration has generated the reproduction of original village structures. Each Turkish pub is a meeting point for village-centred networks. This strong socio-spatial structure creates opportunities for reciprocity. But this reciprocity in turn can be a springboard for the development of survival strategies in the other spheres. Instances of this are support from the network when a member of the community tries to set up an ethnic enterprise, and the temporary employment of someone in order to create his right to unemployment benefits afterwards.

In contrast, deprived neighbourhoods dominated by single persons, one-parent families or atomised ethnic groups are less active in the sphere of reciprocity because of the weakness of their social networks. A great deal of community development can be interpreted as trying to create, regenerate or reinforce reciprocity on the local scale. Moreover, the local is the most obvious spatial scale for reciprocity.

Poor access to goods and services provided by the state in deprived neighbourhoods is fully related to the sphere of redistribution. To a large extent, it results from the strong concentration of immigrants in the poor crescent of the city. This coincidence is explained by the fact that foreigners are still not recognised as full citizens, and are without the right to participate in democratic decisions, even at the local level, notwithstanding the fact that a growing number of them were born in Belgium and many have been present in their communities of residence for a longer time than many Belgian residents. The absence of political rights means that they cannot participate in decisions about what is redistributed and according to which rules. Moreover, without a vote they are easily neglected by local politicians whenever decisions about individual redistribution are to be made (for example, social aid, social housing, individual admission to collective infrastructure, delivery of official forms for social security benefits or grants distributed by regional or federal authorities). The spatial concentration of immigrants adds a collective dimension to their incomplete citizenship. Since they are not represented in municipal councils, their neighbourhoods and all collective infrastructure and services (from schools, sports and cultural facilities to street cleaning and security) are neglected by a political class which has no relations with these neighbourhoods. This neglect further exacerbates deprivation because these neighbourhoods have an inadequate collective infrastructure, inherited from the nineteenth century, when education and leisure time were very limited and state intervention in social matters was seen as a far-fetched idea. Of course, the Belgian population living in the same neighbourhoods experiences the same neglect of the collective infrastructure. Moreover, some poor Belgian households also share the same lack of knowledge and initiative in calling on state redistribution.

The housing problems mentioned earlier are fully part of this problem. Indeed, ever since the creation of a substantial housing market at the time of the industrial revolution, market mechanisms have never succeeded in regulating the provision of decent housing to all households in the city. There has always been a quantitative or a qualitative problem over housing. This market failure, linked to the durable character of housing and to its necessary claim on land, explains state intervention in housing. In contrast to other countries, the Belgian state emphasised home ownership and neglected social housing, which explains the prominence of qualitative (including price) problems in housing in Brussels and more generally in Belgium. This policy also channelled most of the state redistribution in the housing field into the urban peripheries rather than to the poor inner cities.

This general lack of state intervention at different levels is partly tempered by subsidised private initiatives in the fields of health, housing, young people, the elderly, training and employment. Such initiatives open possibilities for survival strategies by poor households, aimed at maximising the aid they receive. But the power and impact of these initiatives should not be overestimated. Poor structural support by the authorities, the specific and complex character of the problems and even party-political considerations are all responsible for the poor results of these initiatives.

It is not only the various state authorities, but also the churches and their charitable actions, that need to be mentioned as redistributors. In some of the ethnic areas mosques play a similar role. In the Turkish area some Turkish shops might contribute to an alternative redistribution system controlled by the mosque. This means that (a part of) their profits are centrally collected by the mosque, which then redistributes these resources amongst those in need of assistance.

Obviously, it would be a hazardous policy to encourage such alternative redistribution systems in deprived neighbourhoods, because, in the long term, they undermine the state's capacity to collect and redistribute enough resources according to democratically decided rules. But deprived neighbourhoods are badly in need of large investments, in both infrastructure and housing improvement. Such improvements, however, would not be appropriate if they resulted in the attraction of fiscally profitable middle-class households, rather than improving the circumstances of the present inhabitants.[15]

Difficult access to the labour market, hindering income attainment through the sale of one's labour, is a problem situated in the sphere of market exchange. Unlike many deprived areas of other European cities, the Brussels poor crescent, just like the deprived areas in other Belgian cities, has the advantage of a central location, allowing access to a large clientele, while the high population densities of the poor neighbourhoods and their concentration of specific ethnic demands open opportunities for ethnic entrepreneurship. Thus inner-city poor neighbourhoods relate in a paradoxical way to market exchange. On the one hand, the growing marginalisation of these neighbourhoods results from the polarisation of the labour market. On the other hand, the inhabitants systematically use (a part of) the market and its mechanisms when they orient themselves to ethnic entrepreneurship or to specific market niches (mostly the cheapest) as strategies to gain access to basic resources.

Ethnic entrepreneurship has a double function as a survival strategy. First, it is a way forward for a significant number of households hit by job loss or unsuccessful in trying to secure employment. The advantages of the socio-spatial features of the poor neighbourhoods of Brussels are such that they foster the arrival in the city of candidates for self-employment from the Limburg mining area and the textile towns in Flanders. And second, these low-budget shops are an important asset for the poor inhabitants in their daily search for cheap products. Interviews with Turkish and Moroccan grocers in the ethnic areas in Belgian cities suggest that the low price setting results from both tough competition between shopkeepers, due to over-investment, and the fact that customers stimulate entrepreneurs to lower their price to the minimum level observed in the neighbourhood. The setting of low prices is thus linked to the poor purchasing power of the neighbourhood considered and the fact that many entrepreneurs are prepared to accept a similar low standard of living.

Despite the general trend that the concentration of poor households generates less market exchange, and where possible more redistribution and/or reciprocity, it also provides an incentive for the development of particular new and cheap niches in the sphere of market exchange. The multiplication of launderettes and telephone shops in deprived areas are instances of this process. Households who cannot afford to buy a washing machine or to repair their existing machine can find an alternative in the launderette, which does not require large amounts of cash for the entrepreneur to establish. For those who cannot afford a telephone, private and relatively cheap telephone shops offer a similar valuable alternative.

Territorial policies against poverty in the market exchange sphere could help ethnic entrepreneurship, especially in niches which are not over-invested in and in those which are suitable for opening up to a clientele larger than the local ethnic group. A more challenging but promising strategy would be to foster light urban industrial enterprises, especially those revitalising the Brussels tradition of luxury goods production, and urban services. But such a policy supposes the removal of the structural barriers in access to employment. Thus efforts must be maintained for the insertion into employment of educationally and professionally marginalised persons, who have had no positive experience of work. But even then, such efforts will never yield more than a few thousand jobs, whereas Brussels has more than 50,000 unemployed persons, not counting those who have not registered as looking for work, many of whom are young people of immigrant origin. Indeed, the labour market of Brussels is determined at the level of the urban region, and its dynamics are linked to the globalisation of the economy and inter-regional competition. Thus, as long as no structural solutions are implemented, the effect of local policies will be too limited. Structural solutions entail the uncoupling of access to a decent income from access to work, through a universal allowance, or a radical reduction of working time in order to give everyone a chance to secure their own resources through the selling of their labour force (Lipietz 1989).

In comparing the three spheres of economic integration, it is clear that none of them is sufficient on its own as a basis for development strategies for deprived neighbourhoods, just as households themselves combine different survival strategies from the three spheres. Indeed, there are no deprived areas which are exclusively focused on market exchange: such a focus would tend to indicate economic success rather than deprivation. But even in the hypothetical case of a deprived market-oriented neighbourhood in which all households chose market-related strategies to escape misery, the overall social tissue would soon be destroyed. Market exchange supposes the pursuing of individual interests by each producer and competition between these actors in the market. Both elements would deepen social inequality in such a neighbourhood and undermine the solidarity and defence of collective interests among the neighbourhood inhabitants, for example in relation to local social development. Similar countereffects appear in strategies confined to the sphere of reciprocity. In such a situation, the poor would end up themselves paying for policies against poverty, while the production of poverty is clearly related to the whole social and economic system and not to the poor themselves. Moreover, conflicts between households capable of functioning in a successful network and socially isolated households would soon emerge and equally undermine their common collective interests. In theory, state redistribution could remedy all this, since the state can guarantee equality and organise a flow of resources from the rich to the poor, or from affluent to deprived areas. Moreover, the state appears, at least for those with full citizenship, as a clearly identifiable and responsible institution against which neighbourhood inhabitants can organise themselves, can negotiate and can require the implementation of agreements. Reality shows, however, that redistribution on its own stigmatises the poor and confirms their poverty rather than fighting against it.

Thus, the best chances for development lie in a balanced use of strategies derived from all three spheres of economic integration. This balance has to be carefully tuned for each type of deprived neighbourhood, since small differences in their socio-spatial features can open or limit various possibilities. This naturally leads to the prioritisation of strategies involving a combination of different spheres of economic integration. For instance, the strengthening of socio-professional integration on the one hand and school-centred community development on the other both imply a reorientation of state redistribution. But they also generate new opportunities in other spheres of economic integration. Socio-professional integration aims at enhancing the chances of young people in the labour market. School-centred community development (including the policy of the ZEP – Education Priority Zones) strives to strengthen the links between schools and their neighbourhoods. The collective involvement of parents in such programmes can turn into new social networks also operating outside the schools and offering access to resources through reciprocity. Social economy combines strategies from the market sphere and the reciprocity sphere. Reciprocity is reinforced when social economy is a substitute for market exchange (for instance in exchange networks dealing with practical knowledge). But in turn, reciprocity can support more efficient action in the market, for example when domestically produced goods are at stake, in the use of consumer durables that can be shared, when difficult access is involved and transport or transaction costs can be shared, or with goods which are cheaper when acquired in large or guaranteed quantities (through collective purchase).

However, this analysis of development strategies in terms of the three spheres of economic integration also shows the limits of territorial-based actions, even when bottom-up survival strategies are included. These limits are set by the necessity for structural solutions, particularly in the fields of market exchange and of redistribution. First, access to labour or at least access to an income allowing a decent standard of living must be secured for every household. This entails a radical reduction of working time or the creation of a universal allowance. Such a situation would also yield opportunities for more efficient redistribution mechanisms, which could make survival strategies unnecessary and lead to real chances for true individual and social development.

Second, a new urban social consensus is needed, in which both the poor and the rich, the young and the old, and Belgians and foreigners decide to put their common interests at the top of the agenda. Within the context of the nineteenth-century belts with their teenagers of immigrant origin, this consensus implies that these young people would at last be recognised as the future of the city who, in the framework of reciprocity between the generations, have the right to training, youth infrastructure, recreation and prospects of decent housing and living conditions. While these rights should be provided locally, they also imply the granting of full citizenship through political rights. Even if many of these young people will eventually acquire these rights through automatic naturalisation (facilitated by changes in the nationality law in 1992 and 1996), the unconditional granting of political rights to all inhabitants of the city after some years of residence is still, as a symbolic gesture, a necessary contribution in the building of this new social consensus.

Notes

1 It is only in the Netherlands that significantly low levels of segregation appear to result from state intervention in the socio-spatial structuring of cities and housing markets (see Van Kempen 1994, and Musterd and Ostendorf in this book).

2 The most accurate and accessible information on these trends can be found in the successive yearbooks on poverty in Belgium (see Vranken and Geldof 1992, 1993, and Vranken, Geldof and Van Menxel 1994, 1995, 1996).

3 The benefits involved do not include health insurance, whose share is increasing. In the period under consideration the total number of unemployed increased by

10 per cent; the retired by 13 per cent for those who had been employees and 8 per cent amongst the former self-employed; beneficiaries of minimum income assistance by 14 per cent among the active population and 43 per cent among the elderly.

4 A self-assessed minimum income 50 per cent higher than the official minimum income level.

5 Besides the minimum income, there is a parallel guaranteed income for the elderly. Recently, the number of recipients has declined slightly because most of the newcomers in the 65+ age category have full pension benefits. In 1996, 104,200 elderly had such an income. Adding them to the minimum income recipients, 1.8 per cent of the population depend for their income on one of these residual schemes of the social security system (not counting their dependent relatives).

6 Adding the guaranteed income for the elderly, 2.5 per cent of the Brussels population depends on a residual benefit (adding in their dependent relatives brings the figure to 3.4 per cent of the Brussels population dependent on such benefits).

7 As discussed elsewhere in this book, social polarisation is widely referred to as linked to the globalisation process and concomitant changes in the division of labour (Sassen 1991). It is difficult to assess this polarisation thesis in Belgium, since measures of social polarisation in relation to the social division of labour are not available. The only income figures available are fiscal income declarations and these are not linked to the socio-economic position of the individuals involved. The only wage figures available are average wages in the secondary sector. They are not broken down by wage levels, by position in the firm, or by skill levels.

8 The index is set at 100 for the first year of the series, but also for 1982 and 1992 because of strong discontinuities in the series. These discontinuities are presumed to have resulted from changes in the definition of taxable income, rather than from changes in earnings.

9 The Brussels urban region defined according to geographical criteria comprises 64 communes. The 19 central communes form the Brussels Capital Region, which is one of the three regions of the kingdom. The central commune of the capital region is Brussels City. In this text, the term Brussels refers to the Brussels Capital Region with its 19 communes and 950,000 inhabitants. The 64 communes are referred to by the term 'urban region', encompassing nearly 1.7 million inhabitants.

10 For a more detailed account of these neighbourhoods and the processes at work, see Mistiaen 1994 and Kesteloot 1995b.

11 The contrast between the inner city and the more affluent suburbs is very sharp in Brussels, but it is also a common feature in Antwerp, Ghent, and to a lesser extent Liège and Charleroi. In the latter two cities the mining and steel industries have resulted in the presence of some working-class neighbourhoods, which today are unemployment zones, outside the inner city.

12 The map shows the 1981 situation, where the spatial concentration of the residual rental sector coincides with the neighbourhoods with the highest proportion of rental dwellings. In 1991 this pattern is somewhat blurred by the growth of the secondary owner-occupied sector (see below).

13 The suggestion that social exclusion and spatial segregation yield highly differentiated spaces of deprivation is fully supported by an inductive analysis of deprived areas in Brussels and Flanders (Kesteloot *et al.* 1996; see also Kesteloot, Mistiaen and De Croly 1997 for a detailed analysis of the Brussels case).

14 The French historian Castel depicts the present social crisis as a process of *désaffiliation sociale* (1995).

15 Most of the financial resources of the Brussels Capital Region and its municipalities are determined by the number of inhabitants and their income. Since the capital

region and most municipalities are losing inhabitants through the income-selective suburbanisation process, the urge to attract middle-class households back to the city is very strong and was articulated in the Regional Development Plan, which aims at recovering 54,000 inhabitants (Vandermotten 1994). Significantly, the planners failed to seize the opportunity to create a large-scale round of public infrastructure investment in the poor crescent of the city.

9

IDEOLOGIES, SOCIAL EXCLUSION AND SPATIAL SEGREGATION IN PARIS

Paul White

Definitional issues

One evolutionary aspect of the social sciences is that it is common for language to be used in a particularly loose way to outline new concepts, with a considerable need emerging later for more precision and tightening up. To take but two recent examples, this has happened with the term 'gentrification' and, more recently, with what has become a catch-phrase of much urban research – 'social polarisation' (Hamnett 1994a).

So-called 'social exclusion' is another concept that joins this group. The phrase has been given official sanction through its use as one of the funding headings for the European Union's Fourth Framework Programme, specifically in that section that deals with 'Targeted Socio-Economic Research', but the definition used is somewhat imprecise. The EU's initial briefing documentation talked of the 'multidimensional processes of social exclusion' and said that it took the form of 'disintegration and fragmentation of social relations and, hence, loss of control thereof. . . . For individuals and particular groups, social exclusion means deprivation or discrimination' (Commission of the European Communities 1994). The document then went on to suggest that social exclusion is particularly caused by employment changes, with demographic changes and associated welfare issues as further significant factors.

The theoretical underpinnings for the association of 'exclusion' with such causal factors are actually relatively weak, and tend to reflect an implicitly functionalist view of social relations. Political processes and cultural understandings are not given any prominence, except through the status quo reference to 'loss of control': interestingly this phrase was replaced in the final work programme by the rather different phrase 'loss of social cohesion' (Commission of the European Communities 1995: 29). The list of shaping factors was extended to include further variables such as lifestyle, education and training, religion, ethnicity and value systems. It is the argument of this chapter that it is the last of these factors – value systems – that is in many ways the most profitable one to pursue in the search for an understanding of social exclusion processes. This inevitably leads into a particular view of social relations and the understanding of society.

Such a view approaches the concept of social exclusion by seeing it as part of the discourse of hegemonic structures, of power and of ideologies within society. To 'exclude' someone or something is to prohibit access to some form of resource: the exclusion mechanisms therefore relate to the interests of those who are instrumental in creating such exclusion for others. Within the social sciences such discourses lead us back to Gramsci's initial formulations of hegemony, the operationalisation of power and the maintenance of social order (Gramsci 1971). Here it is useful to follow Frank Parkin's analysis of 'social closure' through the wielding of power (Parkin 1979), and it should be noted that Parkin specifically talked of 'exclusionary closure' in which dominant social groups are identified by their ability to wield power over others, 'excluding less powerful groups from resources over which dominant groups exert control and to which they have privileged access' (Jackson 1989: 54).

The embedding of concepts of exclusion and social closure within discourses of culture, ideology and hegemony leads into recognition of the fact that the mechanisms of exclusion, the groups so excluded, and the reasons for exclusion, must all be dependent on the particular conjuncture of material, political and societal circumstances at a certain place at a certain point in time. Exclusion is likely to be contested, and will therefore generally be unstable and evolutionary in nature, with societal boundaries shifting in response to tensions built up through existing mechanisms.

The 'resources' referred to by Jackson, over which 'exclusionary closure' operates, can themselves take many forms – jobs, welfare services, education, housing, territory, political legitimacy and citizenship are obvious examples. Similarly, the mechanisms of exclusion include coercive practices backed by the use of judicial powers, the unfettered operation of free-market forces (which could be argued to involve as active a political ideology as coercive or interventionist mechanisms), as well as the more 'conciliatory' imposition of closure through persuasion and through ideology.

To study the dimensions of socially excluded groups, and the mechanisms that are operative, is therefore to attempt to grasp a dynamic and contested reality in which rigid definitions are unhelpful. It is cautionary to remember this, and also to note that the actual collection of data on the socially excluded is always likely to be an incomplete exercise since many of those in this category at any one time are unlikely to appear in official data because their very existence may be subject to legal vulnerability and official marginalisation – with no group is this more true than with clandestine migrants (Winchester and White 1988). There is an obvious need for theorising the existence of social exclusion in metropolitan areas, but when attention is turned to more empirical information we need to be aware that aggregative data-gathering approaches may be woefully inaccurate or even inappropriate, and what is really needed may be 'mixed methods' approaches in which aggregate data are supplemented by the addition of qualitative study from experiential viewpoints.

Ideology and exclusion in Paris

In relating this introductory discussion to the circumstances of exclusion in the Paris region, there is an initial need to identify the hegemonic and ideological structures of society within Paris in order to discuss the groups subject to social exclusion and the relevant mechanisms at work. Here the distinctive administrative circumstances of Paris become crucial. It must be remembered that the metropolitan area of Paris (the so-called *agglomération de Paris* redefined by the statistical office, INSEE, at each census) consisted in 1990 of 398 *communes* or municipalities of which Paris itself was but one (see Figure 9.1). These lay within 8 *départements* of which, once again, the City of Paris was one. The total population of the agglomeration was 9.3 million. Each of the *communes* and *départements* has its own elected authority, operating within highly varied local traditional political cultures.

The overall socio-economic circumstances of the agglomeration are well known: a wealthy core (the City of Paris, with its population of 2.15 million) witnessing a rapid increase in the representation of higher socio-economic

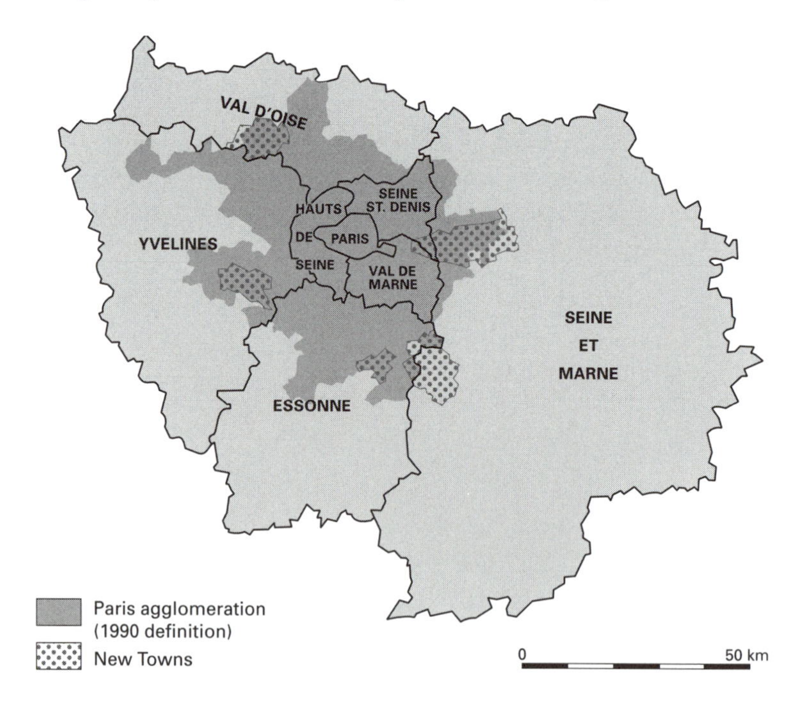

Figure 9.1 Administrative zones of the Ile-de-France region

groups over recent decades; and a generally poorer suburban realm (except towards the south-west), affected by deindustrialisation, but with recently created nodal points of office development associated with the development of five New Towns and other suburban nuclei over the last thirty years.

However, such a description is inevitably superficial. Whilst there are definite elements of a (contested) national French ideology that apply to a greater or lesser extent throughout the agglomeration, there are also distinctive local cultures and ideologies that result in the existence of exclusionary trends in some areas that are absent elsewhere. For example, much of the north-eastern inner suburban realm of the agglomeration still retains elements of the characteristics of the 'Red Belt' of working-class communist ideology that flourished here between the early 1920s and the 1980s (Stovall 1990). Contrasts between local ideologies were particularly apparent at the time of the celebrations of the 1989 bicentennial of the Revolution, when certain areas of the (now gentrified) inner city most associated with the events of 1789–93 were at pains to distance themselves from their past whilst others, particularly in the 'Red Belt', wished to celebrate the triumphs of populist uprising (White 1991).

Whilst accepting the existence of these locally varied ideologies, it is helpful here to draw a broad contrast between the City of Paris itself and the suburban parts of the agglomeration. This contrast is particularly relevant since one of the contested resources in the Paris agglomeration is access to the amenities and facilities of the city itself. It is necessary, however, to start by attempting to describe the national ideology of France today.

A French ideology?

The 1995 French presidential election produced a conflicting series of signals. Apart from traditional class divisions and perturbations brought about by pro- and anti-European Union stances and by candidates standing on ecologist platforms, a distinctive feature was the use of the term 'social exclusion' as one of the key phrases of the campaign. With the right-of-centre candidates making the strong early showing (although the socialist Jospin eventually won the first round of voting, only to lose the second), the ideological agenda took on certain familiar, but also certain new, aspects. The more familiar elements concerned 'immigrants' (the term is used here deliberately) and young people. To these we can add a further ideological flavour through French responses to unemployment. Newer elements emerging in the 1995 campaign concerned fears of marginalisation held by the elderly, and a perceived threat of exclusion amongst non-family households.

If we are to expect any single group to be socially excluded in contemporary France it would be 'immigrants'. Since the start of the rise of Le Pen and the Front National in the early 1980s, discussions of France's immigrant populations have played a prominent role in political debate, in the media, in debates on education, and in public demonstrations. Such discussions have created an atmosphere concerning racism and immigration issues that even the more mainstream parties have had to respond to (Ogden 1987), such that debates on ethnicity inevitably occur within a context that leans towards the vilification of immigrant populations as being implicated in France's high level of unemployment, in street violence and neighbourhood unrest, and in problems within schools.

The French state, however, does not have available to it any particularly strong exclusionary measures to react to such an ideology. Despite some restricted changes introduced in 1993, France still has one of the more liberal naturalisation regimes in Europe. Until the 1993 reform, children born on French soil to 'foreign' parents automatically became French citizens at the age of 18 unless they specifically rescinded that right. Since 1993 such children need to demonstrate their desire to become 'French', but naturalisation is still relatively easy. The vast majority of the 'second generation' are therefore French citizens, and the individuals concerned are lost to data-collection exercises (an example of the problems outlined in the introduction to this chapter). Nevertheless, effective exclusion can operate through other mechanisms, particularly allocation within the housing market controlled by public institutions, which creates strong elements of spatial segregation, as will be discussed later.

It should not, however, be imagined that all foreigner groups are subjected to similar elements of exclusionary ideology (Girard 1977). Antagonism is most reserved for those of North African origin. Other groups, such as the large South-East Asian community, are tolerated (although not necessarily welcomed), whilst groups such as the Portuguese (actually the biggest foreigner nationality in the Paris region in recent years) have effectively 'disappeared' in terms of ideological visibility.

It must also be noted that, although French national ideology can be described as containing an anti-immigrant dimension, it should not therefore be believed that all local ideologies bear the same stamp. This point will be explored further below.

A further important recent aspect of the evolution of French national ideologies has been the move by the right-of-centre government, elected in 1993, to reduce wage levels for young people entering the labour market – the most drastic measure being the authorisation for employers to pay only 80 per cent of the legal minimum wage (known as the 'SMIC'). This element of reduction in what has hitherto been a strongly regulatory approach to labour market control could be argued to be, at least in part, a response to a growing negative image of young people – seen as not being prepared to see their education through to a full conclusion, as hanging out on the streets, as jumping the *métro* barriers, and as being (at least in the eyes of the media) implicated in drug-dealing, petty crime and in a general malaise of fear that keeps many respectable Parisians from using public transport at night.

French social security practices, closely tying entitlements to past employment history and with a fixed time-limit for benefits, have always marginalised the longer-term unemployed and particularly those who have never worked – effectively unemployed youth (Lagrée and Lew-Fai 1989). These are long-standing aspects of French political culture, built into the relatively corporatist structure of the French welfare state (Esping-Andersen 1990). The reduction in minimum wage levels for young people heralded further steps to erode the regulatory nature of the French welfare state, notably resulting in mass demonstrations and strikes in December 1995.

Whilst the growing reluctance to support certain (perceived) groups of young people emerged before the 1995 presidential campaign, the emergence of a strong pensioner lobby in that year was a relatively new phenomenon. The concerns of this lobby were basically two-fold: the first concerned the perceived erosion of living standards through the underfunding of pensions. But the second was more explicitly geographical, and concerned the alleged 'neglect' of districts in which high proportions of the population were retired. In the Paris region, the protesting pensioners particularly drew attention to the long-standing lack of amenities in many suburban areas and related issues of poor accessibility: these issues were clearly contextualised within a more general fear of their neighbours (effectively a fear of the ethnic minority populations of the suburbs) and anxieties about local crime and vandalism. Whether these views have any resonance with pensioners in the provinces is less clear, but certainly they were strongly enough held to bring older people from the Paris region itself onto the streets in late March 1995.

Finally, the strength of the right-wing candidates (Le Pen and the ultra-conservative Philippe de Villiers as well as Balladur and Chirac) led to some anxieties over the emphasis being placed on the significance of the family as a bulwark of French life, and the possible marginalisation of, for example, the divorced and separated, and the stigmatisation of lone-parent households as contributing to social problems.

This discussion has left on one side certain traditional aspects of French political culture such as patriarchy and the depressed status of women. Emphasis has been placed here on contemporary French ideologies that could be instrumental in leading to social exclusion, relating to certain 'immigrant' groups, to the unemployed young, to certain poor elderly, and to those affected by family breakdown. However, as argued earlier, national ideologies can be varied by local circumstances and traditions. Here we shall deal separately with the City of Paris and with its suburbs.

Paris

As is well known, the national politics of France were in large measure played out within the urban arena of Paris throughout the Mitterrand presidency between 1981 and 1995. Throughout his years in power the Socialist president had to cohabit within Paris with the leader of the centre-right RPR, Jacques Chirac, who had been Mayor of Paris since Giscard d'Estaing's presidency (and who was also prime minister for a brief period in the mid-1980s). Mitterrand's view of Paris has been seen as a particularly monarchical one, exemplified in the numerous great projects in which he invested his own and the state's prestige (the Opéra de la Bastille, the Grand Louvre, the new national library, and so on). Chirac's views of Paris were less personalised but more wide-ranging, involving a conscious attempt to establish a complete right-of-centre political and cultural hegemony within the city through the use of area upgrading and planning measures which have resulted in profound changes in the characteristics of the city's housing market.

As Mayor of Paris, Chirac made particular use of two planning measures enacted in 1977 – the SDAU (the Strategic Plan for Development and Planning in Paris) and the POS (the Local Land-Use Plan). These have since been used in ways which have encouraged area-based redevelopment within Paris, with reductions in housing densities and increases in prevalent rent levels. Property development companies have been instrumental in such activities, and the drive for profitability in renewal schemes has contributed to the marginalisation of a particular group among the elderly.

That group consists of tenants living in privately rented accommodation in which the rents are still controlled under legislation introduced in 1948 during the post-war housing crisis. The concept of social exclusion in this case is a very particular one: in relation to the discussion of hegemonic forces and ideology earlier in this chapter these tenants are lacking in power in the face of the dominant forces of property capitalism and a variety of measures, more or less persuasive, have been used to displace them from property that the developers wish to upgrade or to demolish for larger schemes (Carpenter 1994). The rent gap involved with rent-controlled property is considerable: in 1984 it was estimated that controlled rents were at 20 per cent of the free-market level. A planning office survey revealed that in 1989 58 per cent of the tenants in rent-controlled property were over 60, and 37 per cent were over 70 (*Le Monde*, 19 January 1992). These therefore constituted a vulnerable section of the population of the City of Paris.

The conjunction of property capitalism and the organs of the state in social and spatial exclusion can be further demonstrated through two recent legislative features, the first passed under a socialist majority and the second during the brief term of the centre-right parliamentary majority of the period 1986–8. Both Acts have their greatest significance in Paris and are largely an irrelevance in the rest of the country.

Under legislation of July 1985, responsibility for rehousing those displaced by renewal schemes or area upgrading can be discharged by the displacing agency finding accommodation either within the same district (*arrondissement*) of the city, or in an adjacent one – including, where relevant, a neighbouring *commune* of the inner suburbs. Since the areas that the Paris Town Hall was targeting for renewal lay almost exclusively in the poorer outer *arrondissements* of the city, this measure allowed the displaced populations to be removed from the city entirely. Second, under the Méhaignerie Act of 1986 a considerable number of properties with rents controlled under the 1948 legislation had that control removed once they became vacant, even without further improvement, with immediate effects on landlord action and later rent levels.

It has been clear for some time that for the ruling group in the Town Hall the future vision of the residential map of Paris has little space allocated for unwanted immigrant groups (primarily North Africans), for the unemployed, or for the elderly in poor accommodation (in addition to the traditional down-and-outs or *clochards* for whom the City provides night shelters in Nanterre, in the suburbs). Indeed, in the plan for the regeneration of the east of Paris, published in 1987, a specific reason for targeting certain areas was given as their 'ghetto-like' qualities linking high concentrations of foreigners with poor housing conditions while other areas of equally poor housing were left alone (White and Winchester 1991).

This exclusion of the interests of certain groups is therefore subtle: the perceived North African district of the Goutte d'Or is targeted for renewal but not the Chinatown areas of the Arts-et-Métiers quarter or elsewhere. And there has been a notable reluctance to act against the considerable volume of clandestine migrant employment and poor environmental conditions in the Sentier district of the ready-to-wear clothing industry where the use of unregulated (often clandestine) labour is important to the continuing viability of a major economic sector in which there is arguably a convergence between the interests of French capital and of the employed foreign underclass (Montagné-Villette 1990).

Exclusionary ideologies in the City of Paris therefore highlight not only the same groups as national ideologies, but also, at least from the point of view of the Town Hall, emphasise a vision of a 'cleaned-up' Paris from which most marginal groups are removed and in which bourgeois power and lifestyles hold sway.

The suburbs

The opposition of Paris and the suburbs, of Parisians and *banlieusards*, can be seen in dialectic terms in which the two are interdependent but in tension. The trajectories of residential change have been different between the two areas, with a general pattern of *embourgeoisement* in the city paralleling a further proletarianisation of much of the suburban realm. Only in limited areas (towards the edge of the agglomeration but within easy access of the regional express rail links, and in a long-standing sector of the south-west) have affluent middle-class suburban lifestyles really become firmly established.

As pointed out earlier, the suburbs of Paris are administratively fragmented. Ideologies are varied. In middle-class areas such as St Germain-en-Laye or St Cloud, both to the west of Paris, what have been described above as Parisian attitudes hold sway. Elsewhere political cultures are very different. One feature, however, is general throughout the suburbs, and provides a contrast with the

City of Paris. Where in the city there are certain political objectives concerned with actively clearing areas of particular types of resident, in the suburbs such objectives do not generally exist. Instead suburban administrations tend towards defensive and conservative ideologies of preservation (Brechet 1994), or of the prevention of what might be seen as deterioration in existing conditions.

This means that some groups who are becoming socially excluded *in situ* in Paris (primarily elderly private tenants in rent-controlled property) are not regarded in the same way in the suburbs. Although their removal to these areas could, in theory, be seen as a threat to local communities there is, in fact, no negative response. The elderly may increasingly feel isolated and excluded in the suburbs, but the mechanisms at work are not local ones.

The contested ideologies of large areas of the Parisian suburbs are dominated by a different issue – that of the large social housing estates (the *grands ensembles*) and their populations. These estates were built between the 1950s and the 1970s as solutions to a housing crisis in the region as a whole, but from within a few years of their construction many have become labelled as displaying a variety of problems. In recent years their very existence has been problematised, and their tenants have been blamed and vilified for the problems of the suburbs at large. France has been slower to resort to the demolition of such estates than has happened in the USA or in the United Kingdom, although a number of such projects have been carried out in the Paris region (Lelévrier and Pichon-Varin 1995).

As Loïc Wacquant (1993) has pointed out, views of the *grands ensembles* held by outsiders (often including those who live near at hand) are that they consist of immigrant populations heavily skewed towards unemployed youths engaged in a violent and subversive subculture against the norms of French society. This may not be a reflection of reality within the estates, but it is the image and reputation that is important. An implication of such an image is that social exclusion in the suburbs may have different dimensions to it. Particular subclasses of society may be affected by it – the 'immigrant' groups and young people who also feature in national ideologies as marginalised groups – but here there is a further dimension whereby exclusion occurs on a territorial basis ('territorial stigmatisation' as Wacquant calls it) to affect all those of any social group who live in certain 'notorious' *grands ensembles* or, by extension, live in one of the municipalities in which such estates lie. Outsiders, Parisians for example, tend to hold views that deny the social, cultural and political legitimacy not just of those who live in the La Rose des Vents estate (for example), but also of the *commune* in which it lies (Aulnay-sous-Bois), and by further extension (and in conjunction with similar estates and municipalities nearby) the whole of the north-eastern *département* of Seine-St-Denis.

Yet it should be made clear that the residents of the *grands ensembles* are far from being a homogeneous racial minority. Given the common operation of 'threshold of tolerance' concepts in housing allocation policies in France (Grillo 1985), many estates have clear majorities of populations of French ethnic origins, although clustering has been permitted to occur (or been deliberately created) in certain estates (Sporton and White 1989). Not all the estates carry local stigmatisation (Vieillard-Baron 1992), although beyond their own neighbourhoods all estates tend to get lumped together under the same negative labels that have come to view life in an HLM (*habitation à loyer modéré* – the social housing sector) as inferior.

The *grands ensembles* were built at a time when the required scale of new housing provision in the Paris region was beyond the scope of private enterprise. The estates were built for skilled working-class and lower-middle-class residents rather than for the poorest elements in French society. This is a story repeated in many other countries in the early post-war period (Harloe 1995). With the later easing of the housing market these tenants have moved into the growing owner-occupied sector, leaving their social rented apartments to those with less possibility of property purchase – immigrants rehoused from *bidonvilles*, those without stable employment prospects, and single-parent families. The estates as a whole have come to general public attention as a result of a series of disturbances in the early 1980s. It was violence in the Minguettes estate in Lyon that led to Mitterrand's creation of a Commission which suggested target projects for neighbourhood rehabilitation in such estates throughout France (Green and Booth 1996). By 1984 28 such projects had been agreed for the Paris region, 21 of them in *grands ensembles.* Such policies may serve a political and planning purpose, but they also act to highlight, and publicise, 'problem areas'. The media images of the estates and their residents have become nationally known, and reports have tended towards the 'league table' approach of charting estates in the order of their 'problem' or 'no-go' status. Currently around 50 districts in the Paris agglomeration are designated in this way, almost all of them in social housing estates.

Local political responses have been limited in scope. In very few *communes* do the residents of the *grands ensembles* have any real leverage in local politics: in several cases estates straddle *commune* boundaries, often in locations deliberately chosen to produce such fragmentation (Guglielmo and Moulin 1986). Where there are high proportions of immigrant tenants, they lack a political voice anyway since they have not been accorded the vote. Local political ideologies tend therefore to continue to regard the *grands ensembles* and their residents as alien forces, to be contained or ignored. Containment takes the form of municipal lobbying with the social housing authorities to prevent further tenancy allocations of 'undesirable' elements into existing estates: for example, such a campaign was given high prominence in the eastern suburb of Clichy-sous-Bois in 1993. Ideologies of ignoring the *grands ensembles* ghettoise their populations yet further, by denying them any legitimate voice and by pretending that local community life can go on without their participation.

The view from within the *grands ensembles* is a rather different one. Institutionalised social inferiority and immobility apply to all residents, and they are acutely aware of the fact that this exclusion is to a considerable extent operationalised through social housing allocations and ideologies determined by dominant and hegemonic forces. However, this does not bring unity to the *grands ensembles*: instead responses tend to be fragmented, with a variety of strategies used by residents.

One very distinctive feature of the perception of exclusion, however, is that, in direct contravention of wider ideologies that see the *grands ensembles* as locales of racialised violence, within the estates that racialisation is often only very weakly developed amongst the most excluded and vilified elements – the young people. As depicted in a well-known novel and film of estate life, *Le Thé au harem d'Archi Ahmed* (Charef 1983) or *La Haine* (Kassovitz 1995), disadvantaged youth of all ethnic origins act together and where there appear to be racial tensions between young Maghrébins and French tenants these are in reality often generational in nature and reflect exclusionary tendencies against youth cultures as a whole. Young people are therefore, to return to Parkin's term, in a state of 'dual closure' or exclusion, both as *grand ensemble* residents and as young people.

The picture of socially excluded groups in the Paris agglomeration as a whole is therefore a complex and dynamic one, responding both to national and to local political cultures and ideologies. It must be constantly borne in mind that exclusion is a relative concept; and while it may appear on *prima facie* grounds that it applies to whole groups in society, the boundaries are often somewhat vague and fluid. Thus individuals within the potentially excluded groups can sometimes adopt strategies that ensure their continued participation and advancement rather than exclusion.

Location and social exclusion

Location relates to social exclusion in Paris in two ways. Not only are certain of the socially excluded increasingly being displaced (particularly from Paris itself) to more marginal spaces within the agglomeration. In addition, certain such marginal spaces have themselves become the basis for identifications of exclusion through the operation of 'addressism'. In this way social and spatial marginality come together strongly (Vant 1986). The discussion that follows consists of a general overview of the spatial dynamics of the distribution of certain groups over the last inter-censal period (1982–90) at the broad scale of the agglomeration as a whole. This is followed by some brief remarks concerning the meaning of exclusion at a more localised scale, seeking to illustrate the more complex interrelations between marginalising forces, the emergence of underclass zones, and the consequences, both negative and positive, of these developments for those involved.

As discussed earlier in this chapter, exact definitions of relevant groups are not always possible, nor can definitions always be operationalised. Census or other information is often inadequate in dealing with the subtleties of subgroups who may be subjected to exclusion processes. It is only for a limited range of social categories that plausible data are available, and those are not necessarily the most important groups. The true dimensions of homelessness, for example, are not completely ascertainable. Nevertheless, some insight can be gained into the Paris situation by briefly focusing on four groups. These are immigrants; unemployed youth; lone-parent families; and the elderly. It must be stressed that not all people in these categories are marginalised or excluded, but the ideological tendencies discussed earlier suggest that marginalisation processes may be at work.

It is not possible to use the French census for analysis of ethnicity: the data collected instead relate to citizenship or nationality. Given the fact, as mentioned earlier, that France has a relatively liberal law on naturalisation, 'foreigners' in the census can to a large extent be accepted as representing 'immigrants' in the strict sense, plus those currently aged under 18 who have been born in France to such immigrants (and who, until 1993, would automatically become French citizens at their eighteenth birthday) – the missing elements are those who were once immigrants but who have now become French citizens.

Table 9.1 shows the growth rates of the foreign population by major zones of the Ile-de-France region. In this extended area, which effectively covers the Paris Basin as a whole, the rate of growth of foreigners has recently been slower than for the population as a whole. The number of Algerians has been decreasing, partly through the acquisition of French citizenship at the age of 18 by those born in France, although with other demographic processes of mortality and migration also of significance. The numbers of Moroccans and Tunisians have been increasing.

The dynamics of the situation are interesting. From the earlier discussion it might be expected that the importance of foreign residents would be diminishing

Table 9.1 Growth rates, total population and foreigners, 1982–90, Ile-de-France region

	% change 1982–90				
	Total population	*Foreigners*	*Algerians*	*Moroccans*	*Tunisians*
Paris	–1.2	–5.8	–19.8	+9.0	–11.3
Inner suburbs					
Hauts-de-Seine	+0.3	–5.4	–35.6	+3.0	+9.2
Seine-St-Denis	+4.1	+13.3	–19.0	+50.4	+42.3
Val-de-Marne	+1.9	+1.2	–16.4	+48.1	+18.5
Outer areas[a]	+13.4	+9.0	–16.2	+36.1	+13.0
Ile-de-France	+5.8	+2.8	–19.3	+25.7	+7.5

Data sources: Recensement Général de la Population de 1982, Résultats du Sondage au 1/4. Recensement Général de la Population de 1990, Population, Activité, Ménages

Note
a The *départements* of Seine-et-Marne, Yvelines, Essonne and Val d'Oise.

in Chirac's city. That was indeed the case. The total population of the city fell by 25,000 between 1982 and 1990, and 21,000 of that net decline was accounted for by the fall in foreigners. The numbers of Algerians and Tunisians both fell sharply, whilst the number of Moroccans rose, but much more slowly than in the region as a whole.

What is true of the City of Paris is also true of the western inner suburban *département* of Hauts-de-Seine – the richest part of the suburbs and never truly part of the old 'Red Belt'. Here the decline in foreigner numbers contrasted with the overall growth of the population, whilst the fall in the number of Algerians was spectacular. On the other side of the city, the industrial *département* of Seine-Saint-Denis displayed the opposite tendency, with foreigner numbers growing at over three times the rate of change in the total population: thus they progressed from 17.4 per cent of the population in 1982 to 19.0 per cent by 1990. The relationship between foreigners and social housing in this *département* will be considered shortly.

Figure 9.2 displays the pattern of youth unemployment in the agglomeration of Paris in 1990, at the level of the *communes*. Once again, Seine-Saint-Denis stands out as an area of particular interest. Youth unemployment here in 1990 stood at three times the level of the western parts of inner Paris, and although the overall youth unemployment rate in the Ile-de-France fell from 29 to 20 per cent between 1982 and 1990, there were certain *communes* in Seine-Saint-Denis, such as Clichy-sous-Bois and Romainville, where the rate actually rose over the period. North-eastern parts of the City of Paris also showed an above average youth unemployment rate, as did certain industrial areas upstream of the city along the Seine valley. This relationship between youth unemployment and deindustrialisation was also visible in the particularly high unemployment rates of the lower Seine around Mantes-la-Jolie, an area only added to the statistically defined Paris agglomeration in 1990.

It can be added that these first two more socially excluded groups ('immigrants' and the young unemployed) have spatial distributions that are largely coincident, and emphasise the *département* of Seine-Saint-Denis as an area of high levels of representation. This was also true for wider definitions of variables of exclusion: for example, the unemployment of non-European Union citizens (of all working ages) was most marked in 1990 in that *département*, at 23.4 per cent.

The distribution of single-parent families appears at first sight to be very simple and, in many ways, predictable, with a marked gradient from high values in the City of Paris to low values in the outer areas of the Ile-de-France region. In Paris in 1990 22.5 per cent of all families (households with children) had only one resident parent. In the inner ring of suburban *départements* the figure was 17.5 per cent, whilst in the outer ring it was 12.2 per cent. The rate of growth of these proportions between 1982 and 1990 was relatively even throughout the region.

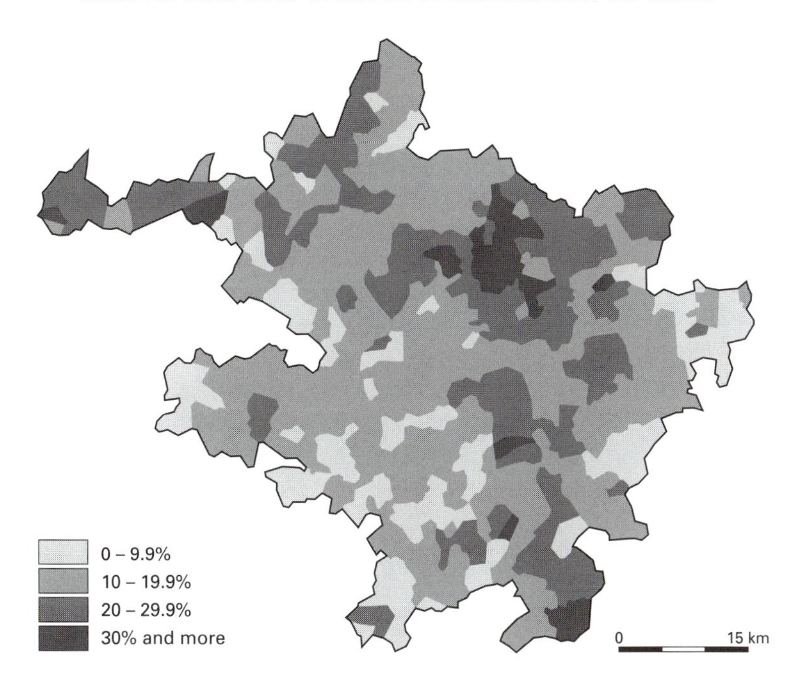

Figure 9.2 Unemployment rate amongst 17–19-year-olds, 1990, by *commune*
Source: Redrawn from the *Atlas des Franciliens*, vol. 2: 125 (INSEE/IAURIF)

However, these figures can be looked at from another angle. The total proportion of familial households is much lower in Paris than elsewhere in the region. Therefore, if instead of relating single-parent families to all families, we relate them to all households, a rather different picture emerges. In these terms the proportion of households made up of single-parent families is actually lower in the City of Paris than in any other *département* of the agglomeration. The highest proportions in fact lie in the *départements* of the inner suburbs (Table 9.2). In addition, the growth rate of the importance of single-parent households has recently been higher in all *départements* of the region than in Paris, with the exception of Yvelines in the south-west. Once again, Seine-Saint-Denis stands out as having the highest representation of this vulnerable group in society, and with the most rapid rate of increase. By 1990 almost one in ten of all households there consisted of a single-parent family.

The final group to be identified here is composed of the elderly. Figure 9.3 shows the proportion of all households in 1990 that had a 'reference person' (effectively household head) aged 60 or over. (There is no satisfactory census indicator for the distribution of the poor elderly.) This distribution is a rather different one from several of the others. The New Town areas towards the outer edge of the agglomeration clearly have low proportions of elderly households, whilst in the western parts of the inner city and in the suburban areas

Table 9.2 Households consisting of single-parent families as a percentage of all households

	1982	*1990*	*Change 1982–90 (%)*
Paris	4.4	6.9	+2.5
Inner suburbs			
Hauts-de-Seine	5.3	8.2	+2.9
Seine-St-Denis	6.1	9.5	+3.4
Val-de-Marne	5.8	8.8	+3.0
Outer areas			
Yvelines	4.8	7.2	+2.4
Seine-et-Marne	4.3	7.0	+2.7
Essonne	5.1	7.8	+2.7
Val d'Oise	5.5	8.1	+2.6

Data source: as Table 9.1

of the south-west over 30 per cent of households are headed by a person over the age of 60. In this dimension the *département* of Seine-Saint-Denis shows up as an area of generally average ageing, rather than being an extreme case (as on the other variables considered).

In looking at the recent dynamics of the distribution of elderly households, recourse must be made to a different variable – that of the proportion of households whose head was retired. This is because the age-categories for the reporting of household heads changed between the 1982 and the 1990 censuses. When the patterns of change are examined on this new variable, however, some interesting findings emerge (see Table 9.3). At the level of the *départements* of the Ile-de-France region, the City of Paris had the highest proportions of retired households in 1982 (22.4 per cent), with certain of the outer parts of the region having markedly lower levels. By 1990 the proportions of retired households had risen everywhere, but with that rate of increase being much slower in the City of Paris than elsewhere: indeed by 1990 two inner suburban *départements* – Hauts-de-Seine to the west and Val-de-Marne to the south-east – both had higher proportions of retired households than did the City of Paris. Certain outer ring *départements* also had rapid rates of increase in retired households, in part reflecting rapid suburbanisation over the past twenty years involving families who were then middle-aged, often purchasing property for the first time.

The discernible spatial patterns from these analyses of recent census data have been considered here at a very broad scale. At the level of individual *communes* much more complexity would emerge. However, a number of general conclusions can be stated from the foregoing discussion.

Perhaps the most important is that the City of Paris was not, during the most recent inter-censal period, witnessing accentuation of the representation of many of the more socially excluded groups in contemporary France. Certainly the problem of homelessness is almost certainly greater in the inner city than

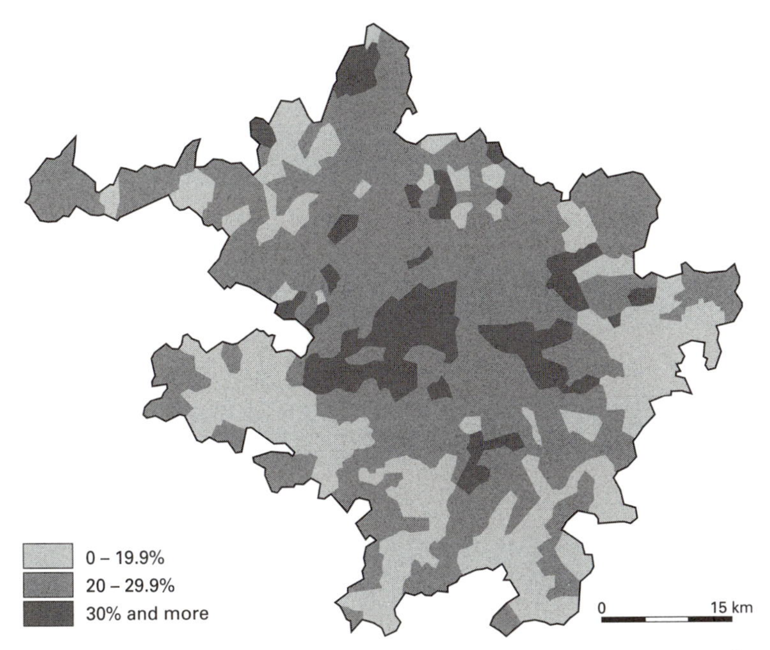

Figure 9.3 The proportion of households headed by someone over 60 years of age, 1990, by *commune*

Source: Redrawn from the *Atlas des Franciliens*, vol. 2: 41 (INSEE/IAURIF)

in the suburbs, but that in part reflects the greater subsistence opportunities of the city centre for a homeless population. However, for the other groups susceptible to social exclusion it would appear that increasing locational concentrations are suburban rather than inner-city in nature. Paris is not keeping up with the rate of growth of immigrants, of single-parent families or of the elderly that is taking place particularly in the inner ring of suburban *départements*.

Table 9.3 Households headed by a retired person as a percentage of all households

	1982	*1990*	*Change 1982–90 (%)*
Paris	22.4	23.4	+1.0
Inner suburbs			
Hauts-de-Seine	20.5	24.0	+3.5
Seine-St-Denis	18.8	22.1	+3.3
Val-de-Marne	19.5	23.7	+4.2
Outer areas			
Yvelines	16.0	20.7	+4.7
Seine-et-Marne	20.2	21.6	+1.4
Essonne	16.4	20.2	+3.8
Val d'Oise	16.7	20.5	+3.8

Data source: as Table 9.1

This is not to say that the City of Paris is not a location for excluded groups. Clearly it has provided a set of housing opportunities for them for many decades. However, during the 1980s the particular politics of planning and the ideology of the Town Hall have had an influence in halting the downward social mobility of many areas within the inner city and of upgrading the habitat and the social composition of the residents.

In this respect Paris presents a very distinctive contrast with many major cities of the developed world, where the trajectories of neighbourhood change in the inner city are more diverse, with gentrifying neighbourhoods contrasting with accentuated marginalisation in other districts in a form of spatial polarisation. Because of the particular right-wing local politics of the city and the fragmentation of the overall administration of the agglomeration, the signs are that during the 1980s the social dynamic of the City of Paris first slowed down and then reversed certain of the forces leading to the concentration of marginalised or potentially excluded groups on its territory.

Today the problems of social exclusion often generally thought of as particularly 'inner-city' in character are actually most clearly demonstrated within the Paris Region in the suburban *département* of Seine-Saint-Denis. Once part of the 'Red Belt', this area has undergone considerable industrial restructuring in recent years and despite the creation of new opportunities through the proximity of the road-based entrepôt of Garonor and the airport-based Roissypole to the north, considerable employment difficulties remain. But the problems of the area also relate to the ideological marginalisation that has overtaken French social housing estates and which has thus had profound effects in Seine-Saint-Denis where in 1990 32 per cent of all households were living in social housing. In that year a total of 59 per cent of the tenants of such housing in the *département* had a head who was normally employed in a manual occupation, was retired, or had no occupational classification. It should not, however, be thought that this marginalisation of social housing relates exclusively to the presence of foreigners: only 18 per cent of Seine-Saint-Denis's social housing households were headed by a foreigner. The social housing estates may increasingly be zones of exclusion, but they are not generally racialised ghettos: French as well as foreigners are excluded in them.

This association of social housing with exclusion is, however, a complex issue since, as discussed earlier in this chapter, the very possession of an address on a *grand ensemble* may be enough to create severe difficulties in obtaining employment, irrespective of the characteristics of the individual (Loinger and Ledoux-Rehoudj 1986). In this way a vicious circle can be set up from which it is difficult to escape – allocations to social housing take place according to need in a situation where property purchase is increasingly becoming the norm; but then residing in a social housing estate results in reducing the possibility of 'earning' one's way out, with difficulties in obtaining employment resulting both from the estate image and from the shortage of jobs within the local labour markets accessible from the estates.

However, it is also useful to consider Wacquant and Wilson's (1993) distinction between different types of marginalised or underclass areas. In their discussion of inner-city ghettos they differentiate between the 'organised ghetto' and the 'hyperghetto'. The first of these has certain positive features for its inhabitants, whilst the experience of the latter is almost wholly negative. It could be argued that the Paris agglomeration has both types of area, but that the organised ghettos of the inner city are now generally in the process of being dismantled through the ideological and planning forces discussed earlier. Such inner-city areas as Belleville and the Goutte d'Or, associated for several decades with high levels of North African immigrants, with high unemployment and with residual populations of the impoverished elderly inhabiting slum housing (White and Winchester 1991), are now being upgraded through clearance policies and through new housing construction, having been labelled in Parisian ideologies as 'undesirable' or 'stigmatised' districts. The important life opportunities offered to their residents through community-based commercial activities (often involving petty trading) and through the presence of important community services are being lost.

In contrast, the 'hyperghettos' of the Paris region today are not in the inner city, but consist of certain of the peripheral social housing estates. Here the concentration of socially excluded populations, combined with the non-existence of any form of economic, social or cultural support, puts their inhabitants at the margins of all aspects of French life today. And in such peripheral estates social exclusion is heavily reinforced by physical exclusion to marginal space within the agglomeration. This is of particular significance since, particularly for those with no formal employment, it is the inner city that can offer most in the way of opportunities for casual earning.

In these peripheral areas mechanisms of social, economic, geographic and political marginalisation all operate together. The social housing estates were created to solve an accommodation crisis for a French employed population of relatively modest means, but forty years on that population is opting for owner-occupation, leaving the estates to house those who have little choice and who consist of those stigmatised by general national and Parisian ideologies. These hyperghettos constitute a major challenge both to French welfare policy and to local administrations, yet there is no sign that any ideological change that would reduce the levels of exclusion and stigmatisation is imminent. Instead there are increasing indications of the inter-generational reproduction of marginality as young people grow up in an environment where they feel a sense of alienation from the rest of French society, and particularly from its hegemonic structures. The results can be seen in a number of local contests over social power, for example through periodic unrest, vandalism and behaviour which is labelled 'antisocial', and which serves both to confirm alternative societal norms within the estates and to reinforce the external images of them as 'no-go' areas inhabited by an underclass.

Discussion

This chapter has argued that social exclusion processes are the products of particular circumstances and are thus specific to periods and places. Whilst the general economic forces at work are undeniably global in their nature, the actual translation of these into local trends depends on historical, cultural, political, demographic and social factors that affect the local arena. Exclusion is created by hegemonic structures and as such relates to social power as much as to economic circumstances. Thus, although the economic status of groups is a key factor in exclusion, it is not the only factor: exclusion also results from 'social constructions' of marginality or vulnerability.

The French context for these representations of ideology is a particularly interesting one, since France has a highly developed welfare state with a particularly important dimension consisting of a substantial social housing sector. In many respects the accentuating spatial patterns of social exclusion now visible around France's bigger cities result from a progressive de-legitimisation of social housing as a housing sector. This itself has two causes: first the operationalisation of desires for suburban home ownership in what has more recently been a freer housing market; and second the progressive association of social housing (and particularly the *grands ensembles*) in the popular mind with problem families, youth unrest, unemployment and immigrants.

The City of Paris is a particularly interesting case, since Paris is one of the very few world cities in which political power is almost entirely concentrated in the hands of the political right, who have then used such power, through planning processes and intervention in the housing market, to promulgate a particular vision of the social nature of the inner city. It is difficult to find a parallel to this situation anywhere else in Europe.

Arguably the hegemonic and exclusionary ideologies operating in Paris (deriving from national as well as local forces) have much in common with those in existence in major cities in other countries. What is different is that the powers have existed within the administrative City of Paris whereby those ideologies can be turned into measurable trends of demographic and societal development. The City of Paris is less and less identifiable as a place of concentration of the more marginal elements in society. Instead the growth of an underclass in the Paris agglomeration is associated with the inner suburbs rather than the inner city, and with areas of large social housing estates in particular.

These developments contribute to an increasing spatial polarisation between the wealthy city (and certain wealthy suburbs, such as to the west and south-west), and the decaying industrial zones of the north-east and of the Seine valley, in which census data suggest that many of the more excluded groups are becoming concentrated. At the same time the niches held in the inner city by certain vulnerable groups, such as the poor elderly in rent-controlled property, or first-generation immigrants in particular *quartiers*, are being broken up by planning intervention. The diverse residential mosaic of the inner city is being homogenised.

The election of President Chirac in May 1995 has not changed these forces: indeed, as Mayor of Paris he was in part responsible for the distinctive social developments within the city and his successor at the Town Hall, Jean Tiberi, has not brought any major change in direction. For at least the years up to the turn of the twenty-first century there seems little prospect of a real reversal in either the ideological exclusion of certain groups in Paris, or their spatial marginalisation to certain areas of the suburban realm and away from an inner city witnessing general social upgrading.

10

SOCIAL INEQUALITY, SEGREGATION AND URBAN CONFLICT

The case of Hamburg

Jürgen Friedrichs

Introduction

All advanced societies exhibit social inequality and, as a spatial outcome of it, residential segregation. Further, the overwhelming evidence from segregation studies shows three concepts emerging as major conditions affecting the extent of social segregation: inequality of income and of education, and discrimination. Of these, income inequality is assumed to have the strongest impact on the spatial distribution of social groups. As has been stated earlier in this book, income inequality and poverty have increased over the last ten to fifteen years, and have led to a discussion about 'social polarisation' (e.g. Mingione 1991: 441, 461), the 'new urban underclass' (e.g. Fainstein 1993, Fainstein and Harloe 1992: 9–13, Kasarda 1990, Kelso 1994, Naroska 1988, Wilson 1990) or, in terms of spatial consequences, about 'divided cities' (e.g. Fainstein, Gordon and Harloe 1992) and the 'dual' or 'quartered city' (e.g. Marcuse 1989).

Two processes account for these changes in highly industrialised societies and large cities in particular: deindustrialisation and social differentiation. With the dramatic changes in the urban employment structure due to processes of deindustrialisation and a trajectory from a goods-producing to an information- and service-based urban economy, the extent of inequality has increased. Two obvious consequences have been an employment–demographic mismatch and an increase in social welfare. The first refers to disparity between the qualifications required for the new jobs in the tertiary sector and the qualifications offered by the local labour force. This has led to rising (structural) unemployment and an increasing number of households depending upon public assistance (e.g. Friedrichs 1985, Kasarda and Friedrichs 1985).

The second process is social differentiation, itself, as Durkheim already observed, a function of the division of labour. Traditional concepts of class or social strata seem inadequate, since to the vertical axis of inequality (as measured by income, education or prestige) a horizontal one has to be added. To capture the diversity of social groups, the concepts of 'lifestyle' and 'lifestyle groups' have been introduced, often with recourse to the work of Bourdieu (1984).

These tendencies are most obvious in large cities. And it is not only in cities which exhibit economic and demographic decline that we can study the consequences of these dramatic changes. Even cities not hit by economic decline or cities in the process of economic recovery also exhibit polarisation, constant or even growing numbers of households with unemployed and on public assistance. The most prominent example is New York (Fainstein, Gordon and Harloe 1992, Lampard 1986), but similar tendencies can be observed in German cities like Frankfurt/Main, Hamburg or Munich.

In this chapter, I will analyse the consequences of these tendencies, using the city-state of Hamburg, Germany, as an example. To organise the presentation of the case study, a theoretical framework will be outlined in the first section. In the second section the framework will be extended to examine the relations between inequality, relative deprivation and social conflict. In the final section, this framework is applied to the city of Hamburg, allowing for more detailed analyses of social and spatial polarisation tendencies.

A multi-level model

It is a well-established fact that social groups are distributed disproportionally over urban space. Less obvious is the process that links (rising) social inequality and poverty to individual behaviour and to spatial structures, segregation in particular. To explore this process, we may conceive of it as a macro–micro problem. I will suggest a set of propositions, and link these to the deindustrialisation process.

Macro level. The basic macro-level assumption is that social inequality has a spatial outcome, residential segregation. To relate this assumption to the micro level requires two problems to be solved:

> The central theoretical problems then come to be two: how the purposive actions of actors combine to bring about system-level behavior, and how those purposive actions in turn are shaped by constraints that result from the behavior of the system.
>
> (Coleman 1986: 1312)

To address these problems, I will first turn to the individual level and then to the context effects. The model resulting from these propositions is graphically presented in Figure 10.1 (the subscript i denotes the independent, j the dependent variable).

Individual level. Most scholars would agree that segregation is a function of income (rent bidding capacity), lifestyle (as measured by the proxy variable

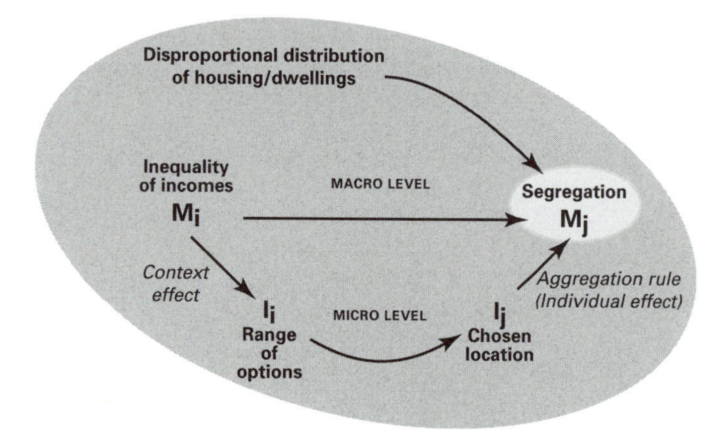

Figure 10.1 A macro–micro model of segregation

'years of schooling') and discrimination, e.g. of ethnic or religious minorities. Income and education are the major resources of the individual or household and the extent to which an individual has these resources determines the number of alternatives available, in this case: residential locations. (The term 'location' is used as an abbreviation to denote both the location and the type of dwelling.) The fewer of these resources an individual or household commands the more they are restricted to a limited range of locations and a single housing market segment. In addition, the degree to which a given minority is discriminated against will further restrict the number of options.

Context effects. These micro-level assumptions can be linked to the macro level in several ways. The most general proposition is that the city provides opportunity structures to individuals and households, the two most important ones being jobs and housing. The job structure is crucial for both the quantity and type of the supply of employment and the educational or qualifications required for jobs. Thus, deindustrialisation can be conceived of as a change in the employment structure comprising three elements: new jobs requiring higher qualifications, making the jobs of semi-skilled workers and part of the skilled labour force obsolete, and increasing the number of jobs in low-skilled services, typically dead-end jobs. The impact of the urban employment structure on the incomes of individuals has been documented in several studies (e.g. Tigges and Tootle 1993).

The second set of propositions pertains to the context effects of the housing market on the households. A basic proposition in segregation theory holds that housing opportunities are disproportionally distributed over urban space. Although this may seem obvious, it is important, since urban planning can affect the distribution of housing by zoning laws or – at least in European cities – by incentives to construct social housing on selected urban sites and thereby influence the spatial market. With respect to housing, deindustrialisation has a negative impact, since lower city tax revenues result in less investment in social housing. Cities experiencing economic stagnation or even decline thus face an extremely difficult situation: while the number of households at poverty levels (mostly due to unemployment) applying for social housing increases, the supply stagnates or even decreases. Moreover, ethnic minority households are more severely hit by this shortage, since they are disproportionally affected by unemployment.

A final set of propositions refers to the context effects of discrimination. From the above assumptions we arrive at a condition of scarcity of jobs and housing. This will lead to fiercer competition between individuals and households for these scarce resources, which in turn increases the probability of ethnic discrimination and conflict.

Relating income, education and discrimination

Let me discuss the propositions of the model in greater detail. The basic macrosociological assumptions accounting for segregation can be drawn from the classic studies of Marshall and Jiobu (1975) and Roof, Van Valey and Spain (1976). They posit that spatial segregation of two social groups increases with:

- the degree of income inequality in a city,
- the degree of inequality of education (years of schooling) in a city,
- the percentage of minorities in the total urban population,
- the size of the total population in a city.

In their studies and in further empirical analyses, all propositions were validated (e.g. Massey 1985: 325, Massey and Denton 1985). The evidence for the first proposition pertaining to the positive effect of income inequality and segregation is inconclusive, since several studies did not confirm such an effect (Hermalin and Farley 1973, Hwang *et al.* 1985, Taeuber and Taeuber 1965). The reason for these inconsistencies may be the interaction of income with either education or ethnic status. Several studies have shown that minority members with incomes similar to those of the majority have less opportunity to move into the residential area of the majority (Erbe 1975, Hermalin and Farley 1973, South and Deane 1993). These results are confirmed by an analysis of segregation in Zurich: the author concludes that the accessibility of areas where native-born Swiss live to members of ethnic minorities is restricted irrespective of the rent-bidding capacity of the foreign-born household (Arend 1984: 34). These findings suggest that immigrants' access to accommodation is restricted by two consecutive filters: the first one ethnic status, a discrimination filter, the second one rent-bidding capacity, an income filter. A similar result is reported by Massey, Condran and Denton (1987). They show that both income and education are crucial variables or constraints on Blacks in their search for housing. Blacks with high incomes do not have access to predominantly White residential areas even if they also have a high level of education. They conclude: 'Upwardly mobile Blacks attempt to move into higher quality neighbourhoods, but are inhibited from doing so by racial prejudice on the part of the Whites' (1987: 32). Further evidence comes from a study of segregation in Rotterdam. The author finds a strong correlation between prejudice against ethnic minorities (Turks, Spaniards, immigrants from Surinam) and residential segregation (Mik 1983).

Hence, discrimination is a strong constraint on minorities in their search for housing, since members of the majority have the power to prevent members of the minority from moving into their residential areas (see Grimes 1993, Hermalin and Farley 1973, Lieberson and Carter 1982, Waldorf 1993). A further constraint pertains to the consequences of the spatial concentration of social groups or strata: their access to the culture of the majority is reduced and, in the case of ethnic minorities, they are caught in the 'ethnic mobility trap' (Wiley 1970).

Discrimination also bars ethnic minorities from the labour market. A recent example of this process is a field experiment by Bovenkerk, Gras and Ramsoedh (1995). In a series of fictitious applications for jobs in a Dutch region, they found immigrants (Moroccans and Surinamese) were rejected in favour of Dutch applicants in 34 to 56 per cent of the cases, depending on the type of job (low or high qualification) and the form of selection (telephone interview, audition), although they had equal qualifications and language skills.

Extending the framework to social conflict

From the above analysis two conclusions can be drawn. First, the analyses of segregation and social exclusion can depart from a multilevel framework using three major constraints: income, education and discrimination. The second conclusion pertains to the explanatory power of education and discrimination. In an urban society of increasing differentiation and diversity of subcultures, both variables may operate not only among ethnic groups and the majority but also between the different majority subcultures. Of these, the variable 'years of schooling' deserves further explication. It is not the sheer number of years of schooling which is relevant; it is rather a proxy for other variables, the – often implicit – assumption being that education is strongly related to behavioural traits which are summarised in the concept of lifestyle. Discrimination is based upon a fear of differences in norms, values and behaviour of social groups other than those one is familiar with.

In such a framework the occurrence of social exclusion and of social conflict would depend upon the extent to which social groups exhibit visible differences in lifestyle. But this is only a necessary and not a sufficient condition for social exclusion. Urban societies have traditionally exhibited a high degree of tolerance or even ignorance of other people's way of life. In addition, cities virtually allow for subcultures to have spatial niches. If, however, economic conditions worsen, the competition of these groups for scarce resources will increase. The most basic resources are housing and jobs. In this competition some groups can become excluded.

However, this analysis seems to be incomplete because we lack the link between income and social conflict. The most promising way to specify this link is by using the assumptions of relative deprivation theory. Although there are variations in conceptualising relative deprivation (see Grofman and Muller 1973), the basic assumption is that a person is relatively deprived if he or she perceives a gap between his or her expected and achieved welfare level. More specifically, we may use the conceptualisation of Merton and Rossi (1968), where the expectation level and the achievement level of an individual are linked to his or her perception of the achievement of persons he compares himself or herself with; as Merton said, 'with people in the same boat'. Relative deprivation is assumed to lead to social conflict or aggressive behaviour (Davies 1962, Gurr 1970).

Relative deprivation thus serves as a link between the changing opportunity structure at the urban (macro) level as perceived by the individual and her or his reaction to this change. It can occur in times of both recession and recovery. In a recession, the individual may ask why he or she has been hit by unemployment while others retained their jobs; during economic recovery the question may be why he or she did not participate in it by getting a (better) job. In both cases, individuals perceive the situation as one in which they do not get what they believe they justifiably deserve. Relative deprivation may lead to social conflict. This proposition will be discussed after the presentation of the Hamburg data.

The case of Hamburg

Hamburg is the second-largest German city and a city-state, i.e. one of the sixteen German Länder. Its dominant economic base has for centuries been shipbuilding, the port, port-related activities and commerce. With the decline of shipbuilding since the 1970s, the city has experienced dramatic losses in jobs in the secondary sector. There has been an increase in the number of jobs in the tertiary sector, although it has not compensated for the losses. The most significant increase is in print-media (e.g. Gruner and Jahr, *Der Spiegel*), making Hamburg the top print-media city in Germany. In addition, there has been a growing number of jobs in media-related activities. As in many other cities, the qualifications required for the new jobs were not met by those laid off from the former secondary sector; hence, a mismatch has occurred, resulting in growing numbers of unemployed and households depending upon public assistance. From the mid-1980s onward, Hamburg fared less well economically than other large German cities, as indicated in Table 10.1.

Table 10.1 Selected economic indicators for five large German cities, 1994

Indicator	*Cologne*	*Frankfurt*	*Hamburg*	*Munich*	*Stuttgart*
Population[a]	963,817	652,412	1,705,872	1,244,676	588,482
% foreign[b]	19.1	29.4	14.6	22.5	23.0
Unemployment rate[c]	12.9	9.2	9.6	6.2	8.4
Public assistance[d]					
Persons per 1,000 inh.	62[e]	64[f]	77[g]	37[h]	27[i]
DM per inhabitant	no data	579	557	221	295
Tax revenue per inh. (DM)[k]	2,051	3,212	6,595	2,422	2,186,92
Debt (DM)[m]					
Total (1,000 DM)	4,999,319	6,472,579	26,069,000	2,636,224	2,216,812
Per resident	5,187	9,921	15,300	2,118	3,767

Notes

a *Statistisches Jahrbuch Deutscher Gemeinden* 82 (1995). Data refer to 31 December.
b *Statistisches Jahrbuch Deutscher Gemeinden* 82 (1995: 26).
c *Amtliche Nachrichten der Bundesanstalt für Arbeit, Jahreszahlen* (1994: 59). Data refer to September.
d Data refer to recipients of 'continuous aid' only.
e Stadt Köln *Statistisches Jahrbuch 1994/95* (1995: 205).
f *Statistisches Jahrbuch der Stadt Frankfurt am Main* (1996: 111).
g *Hamburg in Zahlen* 11 (1996: 339), Statistisches Landesamt Hamburg *Statistisches Taschenbuch 1995* (1995: 504).
h Statistisches Amt der Stadt München *Internet.*
i Stadt Stuttgart *Statistik und Informationsmanagement* (1996: 267).
k *Statistisches Jahrbuch Deutscher Gemeinden* 82 (1995: 394), Statistisches Landesamt Hamburg *Statistisches Taschenbuch* (1995: 185).
m *Statistisches Jahrbuch Deutscher Gemeinden* 82 (1995: 26), Statistisches Landesamt Hamburg *Statistisches Taschenbuch* (1995: 177).

Like all German cities (with the exception of Munich), Hamburg's central city lost population, whereas the suburban zone remained stable and gained by moves from the central city. With the influx of repatriates from Poland and Romania, refugees from the former GDR and migrants seeking political asylum, the population in the central city grew in the second half of the 1980s. In the 1990s, the migration balance turned positive and population losses to the suburbs decreased. A major section of immigrants were foreign-born, leading to a rising percentage of foreign-born. These changes are documented in Table 10.2.

Again, as in other German cities, a stable or slowly growing population is accompanied by a growing number of households. In Hamburg, between 1985 and 1995, population increased by 8.1 per cent, but the number of households by 8.3 per cent, a process mainly due to the increase of one-person households. The economic conditions of Hamburg improved markedly with German reunification. The city, irrespective of its economic problems one of the richest cities in Europe, became a strategically important location for many companies. Both the traditional CBD and an adjacent area to the south experienced a boom in newly constructed office space. As a consequence, unemployment rates declined, but

Table 10.2 Hamburg: selected demographic indicators

Indicator	*1980*	*1985*	*1990*	*1995*
Population Hamburg[a]	1,645,095	1,579,884	1,652,363	1,707,901
Population Hamburg Region	2,618,253	2,582,858	2,672,133	2,988,507[b]
Birth	13,580	12,711	16,693	15,872
Death	23,726	22,266	21,199	20,276
Natural balance	–10,146	–9,555	–4,506	–4,404
In-migration	66,496	56,784	94,215	75,104
Out-migration	64,298	59,792	63,566	68,671
Migration balance	+2,198	–3,008	+30,649	+6,433
% foreign	9.0	9.8	11.5	16.1
Households	808,000	815,800	863,900	881,500[c]
% 1-person households	41.4	44.8	46.2	45.7[c]
Average size	2.0	1.9	1.9	1.9[c]

Sources: Statistisches Landesamt Hamburg (ed.) *Statistisches Taschenbuch* 1981: 44, 184; 1986: 21, 220; 1988: 44; 1992: 23; 1993: 238; 1996: 23, 27, 43, 45; *Hamburg in Zahlen* 7 (1994: 242), 5 (1996: 150), 12 (1996: 400)

Notes
a Data refer to 31 December of respective years; Hamburg Region 1980 and 1985 to 30 June.
b Data refer to 31 December 1994.
c Data refer to April 1995.

demand for housing increased – and welfare expenditure grew as well. These puzzling trends seem to be an inevitable accompaniment of urban economic revitalisation. They are reviewed in detail below. Following the variables outlined in the model in Figure 10.1, I will start with housing conditions.

Housing

As in all major German cities, the housing market has become a major problem in Hamburg (cf. Dangschat 1995). Social housing and housing for lower- and middle-income groups are especially in demand and the need far exceeds the supply. Over the last ten years rent increases have exceeded by far those of wages and salaries. High land values and construction costs have also led to a drastic reduction in social housing. By 1994, the monthly rent for a newly constructed dwelling amounted to a minimum of DM 24 per m^2, but in social housing dwellings tenants are supposed to pay not more than DM 8 per m^2 (both rents exclude heating). Thus, the difference of DM 16 has to be subsidised by the city. Consequently, the construction of new dwellings declined from 1970 to 1990, but has since increased. Even more dramatic has been the low number of new social housing dwellings constructed since 1970. Both factors have aggravated the situation in the housing market. The gap between insufficient supply and a rising number of households demanding social housing or low-rent housing is constantly widening. Data in Table 10.3 document these changes for the 1970–95 period.

Table 10.3 Construction of new dwellings in Hamburg, 1970–95

Indicator	*1970*	*1980*	*1985*	*1990*	*1992*	*1995*
New dwellings (total)	12,087	5,636	4,897	2,826	4,582	9,750
Social housing	4,313	3,549	3,450[a]	1,828[b]	4,109[b]	3,398
Size (in m^2)	68.63[c]	85.9[c]	82.0	90.5	81.6	74.1
Construction cost (DM/m^2)[d]	607	1,406	1,780	1,811	1,926	2,986

Sources: Statistisches Taschenbuch Hamburg 1975: 86; 1980: 97; 1986: 112; 1989: 115, 117; 1996: 109, 100. Personal information by the Federal Statistical Office and the Hamburg Statistical Office

Notes
a New construction only.
b Data for construction permits, increased for one year to allow for comparison with date of completion.
c Residential buildings only.
d Costs at time of construction permit.

Further, the social housing stock is constantly decreasing, for three reasons. First, the financing of social housing is typically done by credits given to a non-profit-making organisation. These credits run for fifteen to twenty years, and are gradually paid back. Large segments of social housing constructed in the 1960s and 1970s are therefore no longer available for social housing. Instead, they are offered on the free market. Second, in 1987 the German government changed a law concerning non-profit-making construction organisations. The result was that a total of 920,000 dwellings in Germany had to be taken out of the social housing stock. Rents, then, are no longer related to costs but to the much higher 'comparative rent' – the rent for equivalent dwellings on the free market. Third, some of the housing associations are paying back their credits earlier than required, thus decreasing the social housing stock more rapidly in order to offer these dwellings on the free rental market.

All these processes have resulted in a housing shortage and higher rents. As a study of the housing market in Northrhein–Westphalia has indicated, annual rent increases are 3 per cent in social housing, 8 per cent on the free market for existing contracts and 21 per cent for new contracts (Veser 1991: 371).

The scarcity of housing has had several negative consequences, both for households and for the city's budget. Tenants seeking a new dwelling (or in need of one) will postpone their move because of the higher rents for new dwellings. It is not overstating the case if we assume that a great majority of existing tenants will, with new contracts, have to pay between 15 and 25 per cent more for their apartments than they do at present. The number of households applying for *Wohngeld* (subsidy for housing) has increased over the last years. The cost of *Wohngeld* comes from the city's budget; in 1991, Hamburg spent DM 200 million for this purpose.

Further, as indicated above, the largest section of social housing is financed by state or local credits over a period of fifteen or twenty-one years. As credits are regressively paid back, rents can be raised by a few per cent. The assumption underlying this method of financing was that rent increases would be lower than tenants' gains in real income. Over a long period this was a valid assumption, but from the late 1980s onward, rent increases have exceeded those in net income. The stock of social housing dwellings dropped from 277,500 in 1987 to 212,108 in 1992 and it is estimated that it will decline to 100,000 by the year 2000 (Freie und Hansestadt Hamburg 1993: 19).

The consequences were manifold. First, the number of households paying 40 or more per cent of their net income for rent increased. Second, Hamburg had to provide additional subsidies for the already subsidised social housing rents. These subsidies amounted to DM 31.8 million (for 50,700 dwellings) in 1985, DM 50.3 million (50,100 dwellings) in 1990 and DM 20 million (27,600 dwellings) in 1993.[1] The reductions in the figures are not due to fewer problems, but to the decreasing number of social housing dwellings. Third, households unwilling to pay the higher rents, or incapable of doing so, were evicted. The number of cases of eviction that went to court was 5,483 in 1991, 5,151 in 1992 and 4,315 in 1993. The number of cases would not have declined if local authorities (*Bezirks-Wohnungssicherung*) had not since 1992 improved contact with tenants, and helped in individual cases to prevent eviction by paying the rent arrears from the communal budget.[2]

Further consequences are documented in the city's report *Poverty in Hamburg* (Freie und Hansestadt Hamburg 1993: 20): low-income households will be excluded from the urban housing market and forced to leave the city, the development of 'islands of poverty' in the north-western, eastern and southern parts of the city, industrial areas with high levels of traffic noise, increasing homelessness among single persons and families too.

The number of homeless has indeed grown in recent years. In 1994, their number was estimated to be 6,000, of whom 3,400 were in communal institutions, 2,000 in small hotels (at DM 35 per day and bed), and the remainder on the street.[3] We thus find two paths to social exclusion: households excluded from affordable housing and among them those excluded completely, either by eviction or being (already) homeless.

Income

Over the past ten years, income in West Germany has increased, from 1990 to 1993 alone by 14.2 per cent, but average gains have been minimal, since the cost of living has increased by 12 per cent (Krause 1994). Moreover, averages conceal processes contributing to social inequality: growing income differences between social groups and growing poverty (11.1 per cent in 1993, taking the 50 per cent poverty line as percentage of mean income). Although households had an overall increase in income, this figure conceals considerable differences. For segments of the population, increases and wages were already below the inflation rate in the 1978 to 1983 period. Since then polarisation of income has continued.

Foreign-born employees have, because of their lower qualifications and hence limited access to higher-paid jobs, lower incomes than German employees. In 1988, the average annual per capita income was DM 35,900 for the foreign-born compared with DM 40,100 for the German male employee; the difference was even greater in Hamburg: DM 34,700 vs. DM 44,200 (Bergmann and Peters 1994: 390).

In the case of Hamburg, two indicators will be used: income differentiation and poverty. Income differences and even polarisation are documented by a recent study of tax revenues in Hamburg (Schüler 1994). The overall increases in annual incomes of tax payers between 1983 and 1989 have been 17.8 per cent in the FRG and 24.3 per cent in Hamburg. Income changes were relatively small in the lower and middle groups (up to DM 75,000 annual income): the number of incomes in these groups as a proportion of all incomes decreased from 64 per cent to 49 per cent; in contrast, annual incomes of households with over DM 1 million increased from 7.4 to 14.7 per cent (Schüler 1994: 117).

Based on tax revenues from income and wages on the level of 104 urban districts, Schüler finds the span of average annual income ranging from DM 31,400 (St Pauli district) to DM 316,250 (Nienstedten district). A classification of urban districts by grouped average incomes is presented in Figure 10.2 (cf. Table 10.5).

The distribution of incomes shows that people in the districts in the west and north, e.g. Nienstedten, Othmarschen, Blankenese and Alsterdorf, have much higher incomes than inhabitants of certain inner-city districts, e.g. St Pauli, Dulsberg, Barmbek-Nord and Hamm. In Table 10.5 data for the 18 lowest- and 18 highest-income districts of the 104 urban districts are displayed.

The income differences increased in the 1987 to 1993 period: in the 'top' district, Nienstedten, average incomes increased by 146 per cent, in the total group of districts above DM 71,000 (excluding Nienstedten) the increase was 36 per cent, whereas in the group of low-income districts (below DM 41,000) the increase was only 9 per cent. In addition, the range of changes was +11 to +145 per cent in the top group, in contrast to +13 to –17.5 per cent in the bottom group of districts. These changes and their ranges clearly indicate that there has been a socio-spatial polarisation in Hamburg.

Unemployment and poverty

Since German reunification, Hamburg has experienced a phase of economic growth, as indicated by declining unemployment figures for both the German and the foreign labour force. However, ethnic minority unemployment rates are still twice as high as those of German residents (Table 10.4). The major reason for this difference is the structure of job qualifications. In 1993, 72 per cent of foreign unemployed had not completed a vocational training, compared with 40 per cent of German unemployed; further, 29 per cent had less than ten years of schooling compared to 8 per cent of the German unemployed

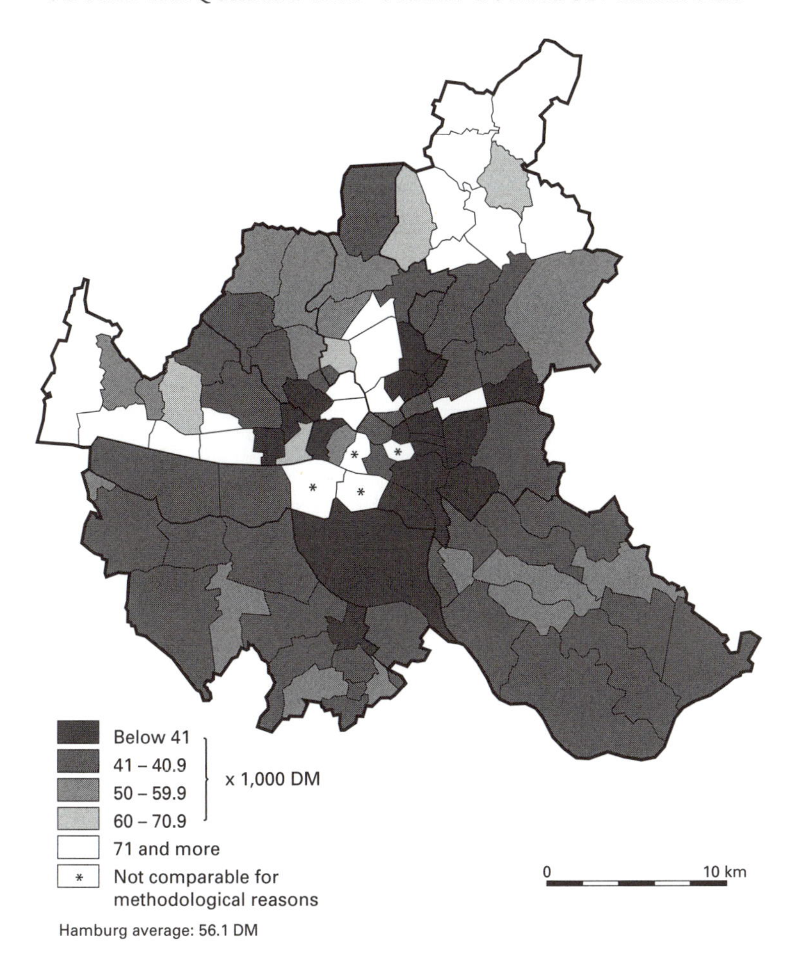

Figure 10.2 Hamburg urban districts by average annual income categories

Source: Schüler (1994: 118, redrawn)

(Freie und Hansestadt Hamburg 1994: 61). As in the past, because of their lower qualifications, minority members will continue to become the victims of economic restructuring and the decreasing number of low-skilled jobs.

The impact of recovery on the level of poverty has been even smaller than on the levels of unemployment. Poverty, if measured by the number and percentage of persons on social welfare, has grown constantly over the last fifteen years. From 1980 to 1990, the number of households receiving constant aid increased from 34,326 to 89,361. Since 1990, the percentage of persons receiving constant aid fell in 1991, but then went up again in 1992.

Again, foreign households were hit more than German households. In less than one decade, the situation worsened dramatically for the foreign population:

Table 10.4 Hamburg: unemployment and public assistance rates, 1988–94

Year	*Unemployment rate*[a]		*Public assistance*[b] *per 1,000 inhabitants*	
	Total	*Foreign*	*Recipients*	*Expenditure (DM)*
1988	12.5	22.1	85	395
1989	11.2	18.5	90	427
1990	9.7	16.0	91	473
1991	8.1	14.6	85	464
1992	7.6	12.9	89	511
1993	8.6	14.6	93	558
1994	9.6	18.0	n/a	n/a

Sources: Statistisches Taschenbuch Hamburg 1989: 81; 1990: 75, 154; 1991: 75, 152; 1992: 73, 150; 1993: 77, 154f.; 1994: 72, 148f.; Bundesanstalt für Arbeit (ed.) *Strukturanalyse* (1994: 67), *Amtliche Nachrichten der Bundesanstalt für Arbeit* 11 (1994: 1700). Personal communication from the Hamburg Statistical Office and the Hamburg Department of Labour, Health and Social Affairs

Notes
a Data refer to 30 September of respective year.
b Continuous aid only.

in 1980, 58 per 1,000 German households received public assistance (all forms), in contrast to 47 per 1,000 foreign households. By 1986, the relations were 85:170, by 1990 the figures changed to 91:249, and declined in 1993 to 77:236. The major reasons for this shift are growing unemployment (in the low-skilled job market segment), in-migration of foreign labour and out-migration of German (central city-employed) residents to the suburbs (Meinert 1993: 11, 15).

Of these reasons, immigration deserves closer attention. From the mid-1980s onward, Germany experienced an influx of repatriates, mainly from Romania and Poland, and a growing number of persons seeking political asylum. While the first group is counted as German, the second is not. Many persons in the first and almost all in the second group depend on social welfare payments. Therefore, the 'foreign population' (more precisely: persons not having German citizenship) comprises two large groups: those who came in the 1950s and 1960s as guestworkers (and their descendants) and those seeking political asylum. The total number of applications for political asylum was 121,318 in 1989, and 438,191 in 1992. It is the latter group that inflates the social welfare statistics for the foreign population, as the following data for the recipients of constant aid (per 1,000 of the German foreign population) show: in 1985, the recipient rate for Germans was 5.9 per cent, for all foreigners 14.4 per cent, but for foreigners excluding those seeking political asylum the figure was 9.1; the respective rates for 1990 were: German 7.2, all foreign 23.7, and 14.3 excluding asylum-seekers (Freie und Hansestadt Hamburg 1995: 42–3).

In sum, by the end of 1993, a total of 183,320 persons received public assistance (all forms), of whom 57,276 were foreign-born; 67,420 persons were unemployed, among them 11,797 foreign-born. The city spent DM 2.2 billion on public assistance, 12.8 per cent of its total budget.

These data mirror the fact that economic growth is no longer related to unemployment and poverty; instead, growth is in large part due to more efficient and less labour-intensive production, the consequences being that growing segments of the population, in particular ethnic minorities, are excluded from the labour market and affordable housing, hence a larger percentage fall below the poverty line.

Socio-spatial disparities

Before moving to the micro level, I will give a more detailed account of the conditions in some of the above-mentioned urban districts. Contrasting low-income and high-income districts reveals marked differences between districts (Table 10.6). Lower-income districts have both a larger range of foreign population and higher rates of unemployment. Judged by the latter indicator, the most distressed areas are Billbrook, Veddel, St Pauli and Wilhelmsburg. They are traditional working-class districts with high population densities, the number of rooms equivalent to or lower than the number of household members.

Assumptions from human ecology would lead us to expect such districts to exhibit greater population fluctuations and a higher crime rate than other districts. The first indicator, the fluctuation rate, is calculated by dividing the sum of in- and out-migrants by the total number of residents. The figure is assumed to be indicative of the degree of social stability and informal social control in a given district. Although the differences between the thirty selected districts are much lower than for other indicators, the fluctuation rate is on the average slightly higher in the low-income districts than in the higher-income ones. However, some of the low-income districts exhibit comparatively high values, such as Billbrook, St Pauli, Veddel and Borgfelde. In contrast, among the high-income districts Volksdorf and Wellingsbüttel have very low rates which can be accounted for by the high number of households of owner-occupied dwellings and single-family houses.

The second assumption pertains to the correlation of both income and fluctuation rates with crime rates. The empirical evidence is mixed. Crime rates are significantly lower in high-income than in low-income districts, but among low-income districts neither income nor fluctuation shows the expected positive correlation. The exceptions are St Pauli and Billbrook. St Pauli contains a traditional red-light district (*Reeperbahn*) and it is therefore not surprising to find high fluctuation and crime rates. Billbrook is a special case due to its small number of residents. In sum, we find higher percentages of unemployed and ethnic minorities in these districts. It is a combination of conditions well known from other studies in German and other European cities.

It is these groups of urban residents we can assume to have the greatest problems in the housing market. The situation, however, is aggravated for

Table 10.5 Average annual incomes of households (taxpayers), in DM, selected Hamburg urban districts, 1989

Urban district	*Income*
Hamburg average	56,075
St Pauli	31,400
Dulsberg	32,609
Barmbek-Nord	35,400
Veddel	35,596
Hamm-Mitte	36,366
Horn	36,442
Altona-Nord	36,467
Borgfelde	36,585
Harburg	36,740
Rothenburgsort	37,062
Hamm-Nord	37,560
Hamm-Süd	38,079
Wilhelmsburg	38,860
Barmbek-Süd	39,412
Billbrook	39,534
Jenfeld	40,320
Eimsbüttel	40,479
Ottensen	40,812
Sasel	71,721
Uhlenhorst	72,905
Poppenbüttel	75,823
Volksdorf	78,597
Duvenstedt	78,597
Rissen	79,809
Gross Flottbek	82,676
Winterhude	83,091
Rotherbaum	84,806
Marienthal	86,732
Harvestehude	92,380
Lehmsahl-Mellingstedt	92,765
Alsterdorf	95,259
Wohldorf-Ohlstedt	100,720
Wellingsbüttel	138,092
Blankenese	172,812
Othmarschen	182,888
Nienstedten	316,250

Source: Schüler (1994)

ethnic minorities, because these households are in addition penalised by discrimination. The segment of the housing market accessible to them is predominantly in deprived areas. But it is in these areas that the lower-class, poorly educated German population makes up a large proportion of residents. Since we know from several studies on social segregation that prejudices against minorities vary inversely with education and social class (Massey, Condran and Denton 1987: 53), we may well assume a high potential for social conflict in these areas.

This assumption is in accord with the relative deprivation hypothesis, especially in the reference group version. An implication of the theory is that persons perceive that the economic position of others is better than their own. This perception, obviously, can appear most readily in urban areas, where the basic economic conditions of the residents are fairly similar. A poor German household may live next to a household seeking political asylum or a household belonging to an ethnic minority (e.g. Turkish) and perceive them faring as well or better in terms of housing or jobs. Since they are foreigners, they are judged to be to blame for this inequality. Persons seeking political asylum may even be envied for receiving public assistance because Germans feel this money should be spent for other purposes, serving the 'German' population.

There is further evidence for this set of assumptions. We may study urban districts with a high percentage of ethnic minorities, but where a larger proportion of the non-foreign population has had a higher education. Since the two groups do not compete for scarce resources, the tolerance of the indigenous group would be higher than in the districts discussed earlier. We should thus observe less inter-group conflict.

Inter-group conflict

This brings me to the final question: the extent of social conflict. Unfortunately, we have no statistics on individual behaviour or aggregated data on incidents of aggression for the Hamburg districts. As a proxy, aggregate crime rates have to be used. These, as data in Table 10.6 show, do not support the hypothesis of the deprived status of a district being related to registered deviant behaviour.

There is, however, another, as yet preliminary, way to test the assumptions about conflict, in a study of the election results for the Hamburg Parliament in 1993 (Friedrichs, Jagodzinski and Dülmer 1994). In this study, we analysed the voting patterns in 104 urban districts in Hamburg. (Ten smaller districts were combined with adjacent ones.)

In a multiple regression analysis three dependent variables were used: per cent non-voting, per cent Green Party and per cent Republikaner, a right-wing extremist party. The results revealed an amazingly simple pattern. Years of schooling and type of dwelling proved to be the best predictors and they account for almost 80 per cent of the total variance. As the data in Table 10.7 indicate, Green Party votes are cast by persons with higher education, while right-wing votes are cast by persons in urban districts with the lowest level of education. So far, this would only confirm the hypothesis that education and prejudice are related negatively. Looking at the spatial distribution of voting patterns for these two extremely different parties, we found disproportionately higher votes for the Greens in districts almost totally separated from those with

Table 10.6 Socio-demographic indicators for selected Hamburg urban districts

Urban district	*Income (DM)*[a]	*% foreign*[b]	*% unemployed*[c]	*Fluctuation*[d]	*Crime-rate*[e]	*Votes*[f]	
						(%) right	*Non-voters*
Hamburg average	56,075	9.3	8.6	0.24	18.375	4.8	30.1
Low income							
St Pauli	31,400	29.9	22.0	0.53	57.691	6.3	47.1
Dulsberg	32,609	15.1	11.9	0.21	12.220	5.2	34.8
Barmbek-Nord	35,400	7.5	8.5	0.20	16.481	4.6	31.9
Veddel	35,596	44.4	13.7	0.36	14.776	9.1	39.8
Hamm-Mitte	36,366	14.8	9.7	0.28	12.860	7.1	36.9
Horn	36,442	13.0	9.9	0.21	13.160	6.3	34.7
Altona-Nord	36,467	19.1	12.1	0.23	28.376	5.6	34.2
Borgfelde	36,585	9.3	11.7	0.32	14.527	5.6	35.2
Harburg	36,740	17.2	10.8	0.29	25.555	5.5	27.2
Rothenburgsort	37,062	17.0	11.8	0.24	17.300	6.1	37.8
Hamm-Nord	37,560	6.1	7.2	0.20	8.689	4.6	39.0
Hamm-Süd	38,079	7.8	8.3	0.21	19.816	7.2	35.2
Wilhelmsburg	38,860	21.0	10.7	0.23	16.548	11.6	35.5
Barmbek-Süd	39,412	9.1	8.9	0.20	17.564	4.1	21.8
Billbrook	39,534	31.6	42.1	1.10	51.718	10.6	44.9
High income							
Volksdorf	78,597	2.8	5.3	0.16	10.526	2.8	17.8
Duvenstedt	78,597	2.5	4.1	0.19	6.551	4.2	22.6
Rissen	79,809	4.9	4.0	0.21	7.489	2.6	22.0
Gross Flottbek	82,676	7.5	5.9	0.20	8.104	2.1	21.1
Winterhude	83,091	8.4	8.3	0.25	13.320	2.9	27.1
Rotherbaum	84,806	14.6	12.1	0.30	28.957	2.6	31.9
Marienthal	86,732	6.2	6.8	0.20	14.864	3.9	26.0

Table 10.6 (continued)

Urban district	*Income (DM)*[a]	*% foreign*[b]	*% unemployed*[c]	*Fluctuation*[d]	*Crime-rate*[e]	*Votes*[f]	
						(%) right	*Non-voters*
Harvestehude	92,380	7.7	8.1	0.21	13.906	2.5	27.4
Lehmsahl-Mellingstedt	92,765	3.0	5.0	0.20	3.089	3.4	19.1
Alsterdorf	95,259	6.3	17.7	0.19	9.583	3.0	27.7
Wohldorf-Ohlstedt	100,720	2.1	5.6	0.26	7.646	3.7	19.1
Wellingsbüttel	138,092	4.1	4.8	0.17	7.723	2.5	18.3
Blankenese	172,812	6.4	6.6	0.22	8.444	2.1	22.3
Othmarschen	182,888	6.4	5.9	0.21	15.732	2.4	21.3
Nienstedten	316,250	6.7	5.9	0.22	10.924	2.0	22.3

Sources: Statistisches Landesamt Hamburg (ed.) (1989) *Statistische Berichte. Ergebnisse der Volkszählung am 25. Mai 1987*; Schüler (1994: 116, 117, 119); Landeskriminalamt Hamburg (ed.) (1992) *Kriminalitätslage Hamburg*: 13–19; Statistisches Landesamt Hamburg (ed.) (1994) Endgültige Ergebnisse der Wahl zur Bürgerschaft am 19.9.1993 (unpublished, prepared for author)

Notes

a Average income of taxpayers, 1989.
b % foreign, Census 1987.
c Unemployment rate, Census 1987.
d Sum of in- and outmigrants divided by number of inhabitants, 1992.
e Number of crimes per 100,000 inhabitants, 1992.
f Elections for Hamburg Parliament, 1993, % votes for right-wing parties and % not voting (excluding postal votes).

Table 10.7 Results of the multiple regression analysis of voting patterns[a] in Hamburg, elections for the city-state Parliament in 1993, coefficients of determination and standardised regression coefficients[b]

Indicator	*Green Party votes*	*Right-wing votes*[c]
Intermediate and High Education	0	–0.69
High school graduates (*Abitur*)	0.89	0
Sixty years and older	–0.45	0
Members per household	–0.69	0.41
Number of rooms	0	–0.42
Rent per capita	–0.47	0
Owners per house	0.57	0
Residential area[d]	–0.45	0
Durbin–Watson Test	3.558	2.912
R^2	84.6%	85.6%
R^2 adjusted	83.6%	85.1%

Source: Friedrichs, Jagodzinski and Dülmer (1994)

Notes
a Excluding postal votes.
b All coefficients are significant at the 0.001 level.
c Parties: Republikaner, Deutsche Volks Union and Nationale Liga.
d % buildings with two or more dwellings.

disproportionately high votes for the right-wing parties. A clear pattern of segregation emerges. These results are presented in Figure 10.3.

Foreign-born residents live predominantly in a few inner-city districts, mostly multi-family buildings from the turn of the century requiring modernisation, or new housing estates (e.g. Steilshoop, Billstedt). Segregation of German and foreign-born, measured by the Duncan and Duncan index, is ID = 34.

Combining information from Figure 10.4 and data in Table 10.6, last columns, we find a variation among urban districts with a high foreign population (14.3 to 41.5 per cent) with respect to voting patterns. While in some of them Greens receive more votes (e.g. St Pauli, Altona-Nord), in others right-wing parties get more votes (e.g. Hamm-Mitte, Veddel, Wilhelmsburg). The major reason for these differences is that the former districts contain a larger number of better-educated residents. This validates the second assumption stated above: the degree of social conflict depends on the extent of prejudice; this, in turn, depends on the extent of competition between social groups.

In general, a larger percentage of low-income district residents either vote for right-wing parties or abstain from voting. The difference between these and high-income districts is significant. In addition, we found a correlation between the percentage of blue-collar residents and the percentage of poorly educated people of nearly $r = 0.90$. In turn, unemployment rate and intermediate and high levels of education are correlated with non-voting ($\beta = 0.37$

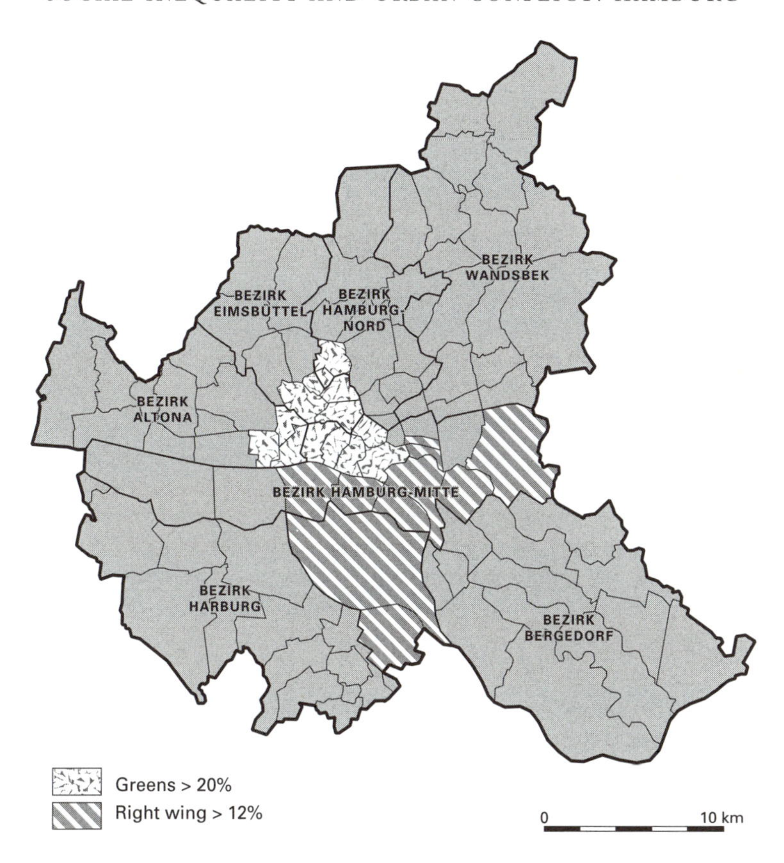

Figure 10.3 Hamburg urban districts with high percentages of votes for the Green Party and the right-wing parties, 1993

Source: 1993 Election data from Hamburg Statistical Office

and $\beta = -0.52$). Thus, we may conclude that in deprived districts residents choose the 'exit' option by migrating, more often than the 'voice' option by voting for extreme parties.

These results allow us to qualify the interpretations of data in Table 10.6 presented earlier. The *relative population diversity* in low-income (or blue-collar) districts may account for lower rates of protest, be they right-wing voting, voting abstinence or deviant behaviour.

Discussion

Segregation is only one indicator of urban fragmentation. As the analysis of Hamburg has shown, polarisation of incomes has, via the housing market, spatial consequences, namely tendencies towards a 'split city'. This is

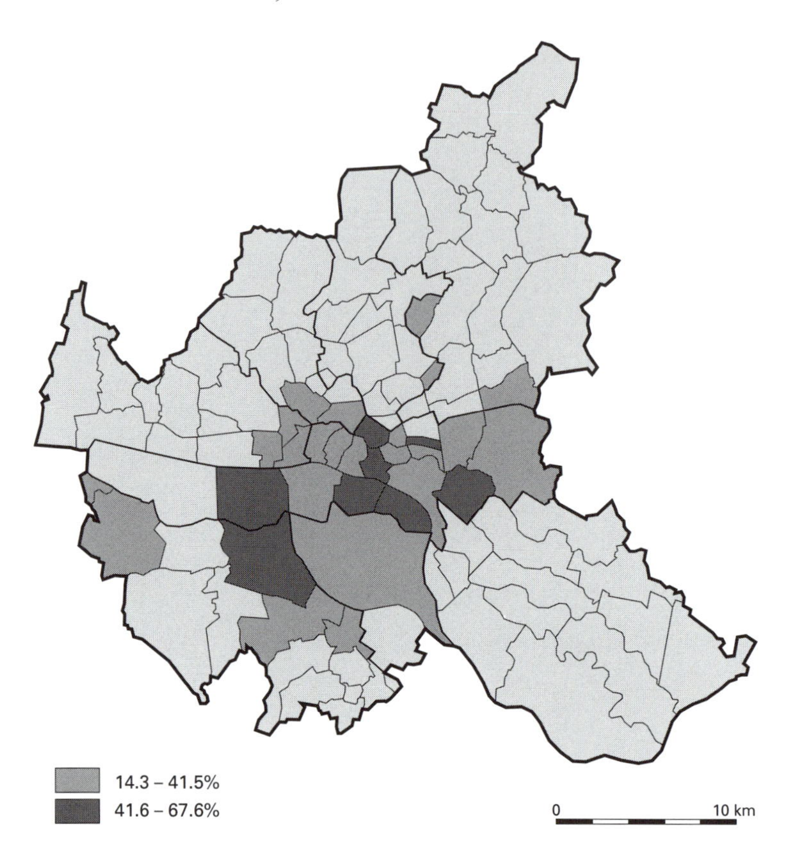

Figure 10.4 Hamburg urban districts with high percentages of foreign-born, 1992
Source: Hamburg in Zahlen 10 (1992: 327)

documented for Hamburg by the concentration of social problems in a small but increasing number of urban districts, characterised by low-income households, substandard housing, relatively high internal densities, high rates of unemployed and households receiving public assistance, higher incidence of evictions from dwellings (for further evidence, see Dangschat 1994).

Thus, the data presented substantiate the major propositions of the macro–micro model suggested in the first section. Processes conveniently described as deindustrialisation or neo-industrialisation can then be defined more precisely as changes in the urban opportunity structure, the (context) effect being to constrain the options of a growing segment of the population, predominantly ethnic minorities. The shrinking urban opportunity structure in both the labour (low skills) and the housing (social housing) markets leads to more intense competition among German and foreign residents. While this alone contributes to a growing prejudice in the German population, discrimination is further raised by German residents' envy of foreign residents for their social benefits. German residents with a low income or below the poverty line will tend to scapegoat foreign residents by blaming them (or the monetary aid they receive) for their distressed situation.

Further, data allow us to explain the paradox of why boom phases are only loosely coupled to a reduction in unemployment and are unrelated to the number of people receiving public assistance. First, economic growth has over the last two decades been achieved by higher productivity, replacing labour by capital (machines). The means by which most companies raised their profit and competitiveness was not an increase but a reduction in jobs.

Second, at least a short boom phase seems to have been counterbalanced by the influx of migrants to a city experiencing economic growth. Judged by the Hamburg data, a larger number of people migrating to the city were ethnic minorities with low qualifications, no income and no other means of financial support than the city they were migrating to, Hamburg in this case. Many of them were seeking political asylum. Since the number of jobs available for people with these qualifications is shrinking (except in the restaurant industry), most of these migrants became or remained unemployed, eventually depending on social welfare. Large cities in highly industrialised countries experience the same problem as their counterparts in less developed countries: migrants imagine there are much better opportunities to get a job than there actually are.

The consequences of this process are presumably similar to those observed in the large cities of less developed countries:

- the creation of informal labour markets, often restricted to one ethnic community (e.g. selling flowers or newspapers in restaurants),
- growing tension between minorities and the indigenous population and among ethnic minorities themselves,
- rising social welfare expenditure in the city, exceeding the increase in tax revenues, thus restricting the city's financial means for investment,
- social exclusion of these minority groups, and spatial separation by the restriction of location to only a few urban areas, typically either new housing estates on the periphery of the city or low-quality inner-city dwellings.

Under the prevailing economic conditions, social conflict among urban groups may rise. One condition contributing to rising conflicts is the cost of a newly rented dwelling (see section 'Housing'), preventing residents from moving and therefore forcing them to stay in a neighbourhood which under other economic circumstances they would have left.

Relative deprivation proves to be a fruitful theoretical link between inequality and segregation on the one hand and social conflict on the other. Contrary to this hypothesis, indications of social conflict are sparse. It seems unjustified to assume social inequality and, more specifically, subjectively perceived relative deprivation by part of the German population will automatically lead to social conflict. The dominant reactions seem to be non-voting or voting for right-wing parties, as our analyses for Hamburg, Berlin and Frankfurt/Main show. (The votes for the extreme right, e.g. Le Pen in France, follow the same pattern.)

However, there are indications of aggression as well. After reunification, in both East and West Germany, brutal acts of aggression against the homes of ethnic minorities occurred, in some incidents (Rostock, Hoyerswerda, both in East Germany) with a large crowd of bystanders sympathising with young people throwing fire bombs into buildings housing families seeking political asylum. Initially, these incidents were limited to cities in the former GDR and were accounted for by their inexperience with ethnic minorities, high unemployment rates and declining economic conditions for a greater section of the population. But only weeks later similar acts of aggression occurred in West Germany, e.g. Solingen.

Overt conflict is especially likely if economic instability or even recession continues. And since instability and only a short phase of economic growth seems to be the more probable outlook for the years to come, the living conditions of ethnic minority members with low qualifications will become very difficult. They, in turn, do not have the 'exit' option as long as political and economic conditions in their home countries do not improve. Moreover, the majority of them have lived in Germany for ten years or more, many as third-generation residents, and Germany has *de facto* become their (new) home country. It is the third generation of the former 'guestworkers' who are caught in the dilemma of being both estranged from their home country and socially and spatially excluded in Germany. Depending on the degree to which the members of an ethnic group are excluded, their reaction may well be to regain their identity by a re-ethnisation and disappointed withdrawal from assimilative behaviour.

Notes

1 Personal information by Jörn Köster, Baubehörde, Amt für Wohnungswesen.
2 Data from personal communication by Roland Günther, Behörde für Arbeit, Gesundheit und Soziales, Hamburg.
3 Personal communication by Roland Günther, see note 2.

11

SEGREGATION AND SOCIAL PARTICIPATION IN A WELFARE STATE

The case of Amsterdam

Sako Musterd and Wim Ostendorf

Social exclusion, social participation

Today, social exclusion, which we will define as related to a lack of social participation, is clearly one of the most threatening problems for the authorities of cities. Moreover, social exclusion will also be one of the key issues cities will have to deal with tomorrow. Disintegration and fragmentation of social relations may function as seedbeds for inter-group tension and criminal behaviour, and since social exclusion implies living in a more or less isolated situation, cut off from mainstream society, the opportunities for social mobility, through additional education, getting a (new) job or an improved housing situation, will be diminished. This in turn may further add to problems of deprivation.

Of course these problems are not new, but during the 1980s the social problems many western cities were facing seemed to increase. Alarm signals were sent out continuously, particularly from the United States and Great Britain. Many researchers and politicians pointed at an increase in social polarisation between often well-educated people included in the labour market and society and – often under-educated – people excluded from that market and society, or between people with and people without a well-paid job. Social conflicts were more visible, sometimes even ending in urban riots (South Central, Los Angeles; Brixton, London, etc.). Many people increasingly avoid outdoor activities and such lack of participation is no longer confined to the elderly. There is also a clear increase in the number of supporters of extreme right-wing parties or bodies, and this also appears to be indicative of an increase in social problems. In addition, unlike the United States, where concentrations of ethnic and poor population groups in cities are more or less common features, most European states have only relatively recently been confronted with the development of areas in which ethnic groups or poverty dominate the picture. The likelihood that neighbourhoods and their inhabitants have mutually negative effects on each other may increase as these concentrations get higher. In addition to these factors, there are many indications that the number of people living below the poverty line has increased rapidly and homelessness seems to have skyrocketed. In short, many interrelated social problems seem to have become bigger and bigger (Wilson 1987, Short 1989, See 1991, Fainstein and Harloe 1992, Marcuse 1993, Jencks 1994).

In order to understand the process of social exclusion and the decline of participation in society, several interrelated explanatory dimensions must be taken into account. The changing economic structure, the developments in the sphere of international migration and the character and restructuring of the welfare state in its broadest meaning, as well as the segregation of the population itself, are the most relevant dimensions in our view. In this contribution we will highlight the influence of the welfare state on social polarisation and spatial segregation and we will also stress the effects of a concentration of relative poverty, or a concentration of those with little prospect of improving their situation, on the level of participation in society. So, in the next section we will elaborate on some notions which are thought to be important to explain (the rise of) social exclusion. Attention will be given to socio-economic, labour market and cultural dimensions, but we will accentuate the expected importance of these dimensions within the framework of the specific characteristics of the Dutch welfare state, compared to some other welfare state models and that of the United States in particular (in other work we present a deeper analysis of the Dutch welfare state: Musterd and Ostendorf 1995). In a later section special attention will be given to some disputes about the effects of segregation on social exclusion or social participation, again within the context of the Dutch welfare state. Second, in the fourth section, we will provide data on the homogeneity of the population with respect to income and ethnicity in Dutch cities and urban regions; to what degree is the situation in Dutch cities to be described as segregated and/or socially or ethnically polarised? In the fifth section we will use some empirical information to test the ideas formulated. Attention will be focused on data that illustrate the effects of the spatial concentration of poverty on the social participation of the population (we prefer to apply the concept of social participation rather than the concept of social exclusion in these empirical parts of the contribution). Personal characteristics and changes in society are of course also most important to the opportunities to realise social mobility (the quality of life). But here the question is whether a geographical concentration of more or less excluded groups, living in deteriorating and devalued areas, adds an extra negative effect to an individual's social situation. In other words, the basic question dealt with in this contribution is: *does the (segregated) neighbourhood make a contribution of its own, reducing social participation (and through that reducing all other social inclusion opportunities), after controlling for personal characteristics, even in cities in a welfare state such as the Dutch?*

Some conclusions will be drawn in the final section, and some expectations will be presented as far as segregation and social participation are concerned, related to the conceived need to restructure the Dutch welfare state.

Social exclusion and the Dutch welfare state: some notions

Social exclusion is one of the most important potential outcomes of many of the processes related to the social problems outlined above. To be more specific, problems related to income (the poor), housing (the homeless), labour (the jobless), education (the unskilled), demography (the elderly) and culture (the non-White) are all potentially firmly linked to social exclusion. People in the categories mentioned are often regarded as excluded; the others as included. Exclusion/inclusion is strongly related to social participation and that relationship is doublesided. A low income will prevent a household from participating in all sorts of activities that cost money. That will add to the isolation or exclusion of the household involved. The same holds true for homeless people. Also, irrespective of the financial consequences, people without a job are not only excluded from the labour market, but from many social contacts, which are related to having a job. In addition, elderly people tend to live in a more isolated situation. Apart from their income and job situation, this may be linked to their physical condition. Their vulnerability makes them easy victims of criminal assaults. Many former relatives and friends may have passed away with old age, implying a reduction in the pool of social relations. Finally, there may be cultural or discriminatory reasons for not participating in mainstream society. Once someone does not fully participate in society, the likelihood of being included at a later stage in time is limited.

Most experience in this field, however, originates in the Anglo-Saxon world. Some researchers have expressed their doubts about the direction of these social processes, and that critique deserves attention, if only to put the Anglo-Saxon experience into proper perspective. One of the most critical commentaries has been formulated by Hamnett (1994a, also in chapter 2 of this volume). He firmly criticised Saskia Sassen's work, in which she argued that social polarisation in global cities, such as London, was increasing during the 1980s. Hamnett showed empirically that this was not true in London, as far as the professional class was concerned. He argued that professionalisation, not polarisation, was the dominant process.

Other doubts about the type, speed and direction of the processes related to social exclusion or participation arise from looking at cities in other European countries and the Netherlands in particular. Research has shown that the historical, ethnic–cultural and state–political differences between the Anglo-Saxon and the Dutch 'worlds' are numerous and large (Esping-Andersen 1990). These differences have contributed to the existence of less polarised urban societies in the Netherlands, compared to cities in, for example, the United States.

Detailed comparison shows that social inequalities between population categories are much smaller in Dutch cities compared to US cities. The state has always played an important (redistributing) role in relation to consumption. The welfare benefit system is extensive. People without a job and also the ill and disabled are firmly supported in terms of income and remain 'included'. Moreover, everyone can lay a claim to and has access to housing. A large decommodified social housing sector, together with a well-developed allocation system, are available to meet housing needs. The size of the sector enables allocation to different socio-economic categories of households and this undoubtedly has contributed to the avoidance of sharp social and spatial segregation. The Dutch thus seem to have realised a situation in which poverty in terms of consumption (income and housing) has been avoided. The chance of becoming excluded socially because of having insufficient income or living in inadequate housing has been reduced.

Quite another story will have to be told with respect to position in the labour market in the Netherlands. Though the loss of a job does not immediately imply a hopeless situation in terms of income and housing, the loss of social contacts made at work, and the lack of prospects of social mobility, are still very important factors contributing to social exclusion processes and new (modern) urban poverty. Since unemployment and semi-unemployment (unemployment hidden in the disability benefit programme) are high in the Netherlands, particularly in the cities, this problem of being socially immobile is a severe one, especially for low-skilled people. The ethnic component contributing to social exclusion differs somewhat between ethnic groups, related to the variation in cultural distances between autochthonous Dutch and Surinamese Dutch (small distance) and the Dutch and Mediterranean population (larger distance).

In short, in Dutch cities the chance of becoming excluded socially will be only moderately affected by factors such as income and housing, since a sustained consumption level is guaranteed in Dutch society. The labour market situation and ethnic–cultural factors appear to be more important.

Segregation and social participation

We started this contribution by stressing that we would pay special attention to the analysis of the (lack of) effects of a geographical concentration of more or less deprived households on someone's chances of being excluded. Of course such an analysis cannot be carried out without controlling for other relevant dimensions: income, housing, labour and education, ethnicity and demography.

There are at least two reasons to be sceptical about the importance of a neighbourhood effect (a spatial segregation effect) on social participation in the Dutch situation. The first reason has to do with the level of segregation that is reached in Dutch cities. Massey and Denton (1993) have experimentally shown what the negative effects of a strong concentration of poverty may be (also chapter 4 in this book). But in the context of the Dutch welfare state model such strong concentrations have hardly developed anywhere, although recently income segregation has increased slightly. However, a concept such as 'hypersegregation' is without value in the Dutch case. We have already mentioned the smoothing effects of income and – particularly – housing redistribution processes. As a result, we assume, the level of spatial segregation of the population, of the concentration of poverty, of people with few prospects of social mobility, will simply be insufficiently high to produce a negative spiral movement, or a so-called 'cycle of poverty'. In agreement with Massey and Denton, we expect that only in situations with relatively high concentrations of poverty will there be over-proportional effects of macro developments, such as economic restructuring, that may prove to be very negative for certain categories of the population (especially unskilled traditional workers). Even though politicians and administrators are extremely afraid of such effects, the concentration of poverty will, in the Dutch situation, be too small for the losers to become the major group in a locality.

The second reason is unrelated to the first. A debate is also going on about the effect of segregation on social problems in city neighbourhoods irrespective of the level of concentration. Influential scholars, such as the sociologist Julius Wilson (1987), stress that living in a ghetto leads to a higher propensity of becoming socially excluded. Robson (1988: 7) shares his opinion; he states: 'The context within which a household lives and the personal characteristics of its members are therefore important additive contributors to the chances of that household's well-being. Place, as well as people, matters.' De Lannoy and Kesteloot (1990: 143–4) underline these remarks:

> Residential differentiation is not only the result of class relations in society, but also the cause of the reproduction of the class relations. Living in a neighbourhood has important effects on the access to collective services and goods and on the norms, values and behaviour of the inhabitants. Residents in deprived neighbourhoods face reduced opportunities for upward social mobility, while there is a good chance that their situation of poverty will be handed down from generation to generation [translation by the authors].

Others, however, state that many researchers assume a causal neighbourhood effect too hastily (e.g. Hamnett 1979, Badcock 1984). Often the association between neighbourhood quality and the problem score will be spurious and explained by the characteristics of the population itself. Their view is that causes have to be looked for in society first. Only after that has been done may a modest extra neighbourhood effect be found as well. Vos (1997) points out that such compositional effects of the residential environment certainly can exist, but that a careful analysis has to be carried out before one can come to a positive conclusion.

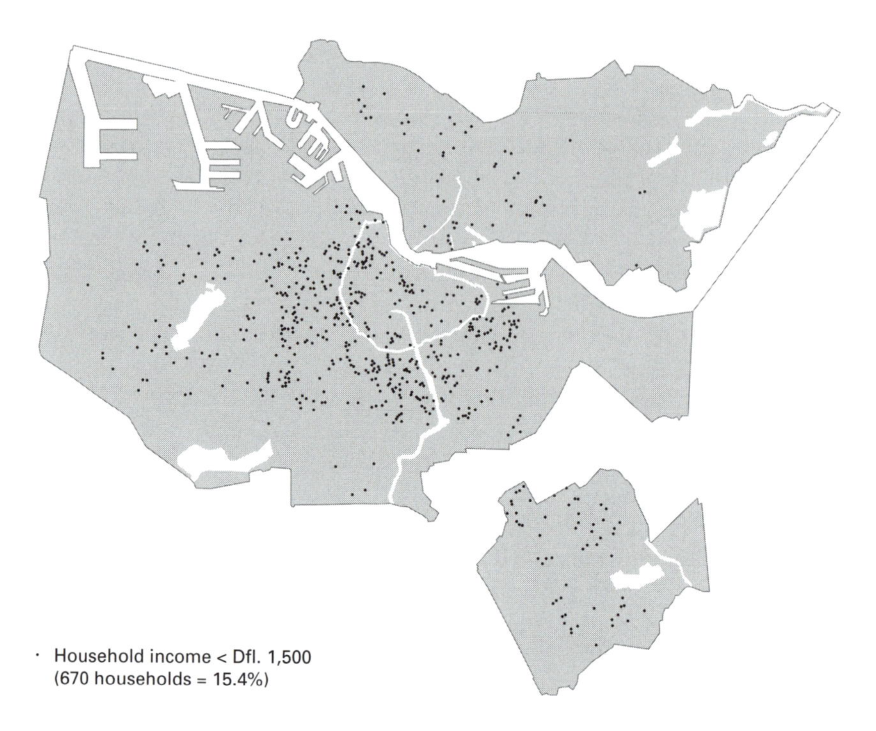

Figure 11.1 a The spatial distribution of households according to household income, 1994 (low-income households)

Source: Surveys AME, University of Amsterdam

This discussion shows that there is no agreement over neighbourhood effects or spatial segregation effects on social participation. For the moment a good hypothesis seems to be that, in the Dutch situation, social participation will not be dramatically affected by what are small differences between neighbourhoods in terms of concentrations of deprived people.

Income segregation and ethnic segregation

In Dutch cities many spatial patterns can still be labelled as 'mosaics' instead of 'polarised entities'. Spatial differences between households with different incomes or ethnic origins are relatively small, although there is some variation depending on the level at which analysis is carried out. As far as the income differentiation is concerned, the suburbanisation process of the 1960s and 1970s was most important. High- and median-income groups were over-represented in the large outward movement from the cities during these decades. The result was a clear – though not absolute – socio-economic gap between the central city and the surrounding suburbs (Table 11.1). Although that gap

Figure 11.1 b The spatial distribution of households according to household income, 1994 (high-income households)

Source: Surveys AME, University of Amsterdam

has changed over time, there has been a tendency for an increase in the city–suburb income difference (Sociaal en Cultureel Planbureau 1996). At the local level, within the city, income segregation appears to be relatively stable over the decades. Figure 11.1a and 11.1b provides some information of the intra-urban spatial distribution of households according to their income. An immediate impression from these figures is that there is hardly any difference between the spatial patterns of the top and bottom income categories in the

Table 11.1 Mean household income (standardised figures) in the four big cities and the urban regions, 1977–94

	1977	*1985*	*1994*
Four big cities (a)	17.5	21.0	27.2
Urban regions (b)	19.5	24.9	34.2
The Netherlands	17.7	22.7	30.7
(a)/(b)	0.89	0.84	0.80

Source: Sociaal en Cultureel Planbureau (1996: 197)

Dutch capital. Homogeneous neighbourhoods can only be found at a very detailed spatial level.

Ethnic segregation, however, is more pronounced at a regional level. Most immigrants to the Netherlands appeared to be attracted by, or become associated with, central city locations. Some 40 per cent of households in Amsterdam are immigrants and about 70 per cent of these originate from non-western countries (labour migrants, refugees). In the Netherlands, the suburbs are almost completely White. One exception is formed by some of the new towns. A close connection with the 'donor-central city' and sometimes a high proportion of social housing have resulted in ethnic minority households moving to these areas too.

However, in this contribution we will focus on segregation within the central city in particular. This indicates that, as with the income patterns, only moderate ethnic segregation patterns have developed (Figure 11.2). Even on this very small scale (1,200 cells with on average 500 inhabitants), the Index of Segregation (Turks and Moroccans versus the rest of the population) measured in 1994 was only 47 (Daalen, Deurloo and Musterd 1995).

From other research we know that during the 1980s and early 1990s segregation did not increase significantly, either in terms of incomes, or in terms of ethnicity (Musterd and Ostendorf 1996). As far as incomes are concerned, during the 1980s and early 1990s, the inner city strengthened its position compared to other areas in the urban region. The inner city changed from an area with a below-average income to an area with an above-average income. This resulted in a more even spatial distribution according to income per earner. The same holds true for the residential pattern of ethnic groups: no increase in segregation occurred. During the 1980s the immigrant population, until then migrant workers residing in lodging houses, started to be reunited with their families and to gain access to family houses through the municipal housing allocation system. The concentration of ethnic minority groups in central parts of the city therefore decreased, while it increased in outlying neighbourhoods with a relatively cheap housing stock. In this respect the result is a more or less stable level of segregation.

To state that there is no concentration of ethnic groups at all is too rigid, but it is quite clear that the fast increase of immigrants entering Amsterdam during the past two decades did not lead to extreme segregation or polarisation. An important factor explaining the low levels of segregation can be found in the organisation of the Dutch welfare state, including special arrangements in terms of income redistribution, social housing, social security, subsidies, benefits, and so on.

Segregation and social participation in Amsterdam

Within the framework set out so far we will now present some data to investigate the hypothesis that in the Dutch situation the spatial segregation of

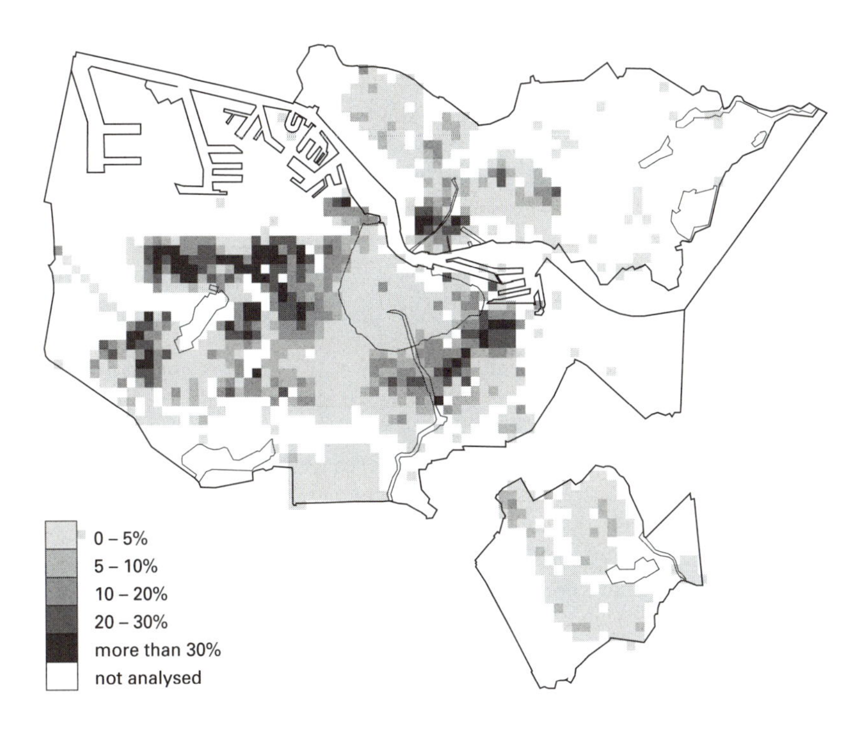

Figure 11.2 The spatial distribution of Turks and Moroccans for a grid area in Amsterdam, 1994

Source: Municipality of Amsterdam

poverty has limited or no influence on the social participation of the population, after controlling for the factors we have already dealt with in this chapter.

Often the instruments measuring poverty or relative deprivation combine income and employment data. From the discussion above, it appears to be very important to separate these two, especially in the Netherlands. This was not done in recent research projects aimed at understanding social participation, which were carried out in the Netherlands by the Dutch Social and Cultural Planning Agency (Sociaal en Cultureel Planbureau 1992) and by the municipality of Amsterdam (1993). This could be the main reason why these projects showed no significant relation between the concentration of deprivation and social participation. Consequently, in this study another indicator of poverty or relative deprivation is used: a person is considered to have little prospect of upward social mobility if he/she is low-skilled and unemployed or in the Dutch disability-programme.

The data used to test the hypotheses were derived from a large survey carried out in Amsterdam in 1990. About 4,000 face-to-face interviews from a random sample of the population of 18 years and older were available for analysis. The data contain many social, demographic and economic characteristics at the individual and household level, as well as some information on the lifestyles of the individuals. The location of each respondent is known in detail since the postcode was added to the records. This offered us the opportunity to use a sophisticated way of measuring the effect of spatial segregation of poverty. The segregation score for each individual was calculated as the percentage of people without prospects of upward social mobility (i.e. being low-skilled, unemployed, or receiving disability benefits), not including the individual respondent, in an area with a radius of 500 metres from the individual (Figure 11.3).[1] Four categories of segregation were distinguished, varying from 4 per cent of people without good prospects surrounding the individual respondent (which was the case for 34 per cent of all respondents), to over 12 per cent of people without good prospects (true for almost 10 per cent of all respondents). Again, these figures show a modest level of segregation. Obviously, these differences in concentration result in different probabilities that individual respondents themselves will suffer from lack of prospects for social mobility: the probability is higher in an area with many people who lack such prospects, and lower in an area with few people who lack prospects of social mobility. This complicated relation between having few prospects themselves as well as being in a concentration of people with few prospects, and social participation, will be dealt with later on.

Ideally, the indicator of social participation must be associated with some measure of the lack of participation in mainstream society. The available data allowed such lack of participation to be measured in two ways. The first was by counting the frequency of staying at home during the evening. In the survey the question was asked: 'How many evenings a week do you usually spend at home?' The second indicator relates to the frequency of visits to pubs, discos, dance halls, restaurants, other dining places, cinemas and theatres (participation in society for which money has to be spent). We also analysed sports activities. An index was constructed on the basis of the answers to the questions on these kinds of participation.

These indicators were also used by the Dutch Social and Cultural Planning Agency to denote social participation in society. However, they are certainly not ideal for measuring participation in a broad sense. They may be criticised for their 'yuppie' character, and we agree that analysis of contact networks with family, friends and neighbours would offer relevant insights, as well as subjective measures such as personal feelings of loneliness. Unfortunately, these indicators were not available in the dataset we were able to use. However, other sources were available for investigating the inter-relations between the indicators used here and other relevant indicators of social participation. This points to a fair correlation between evenings spent at home, degree of social participation and feelings of social belonging.

For a good understanding it is stressed that no indicators such as unemployment or 'on benefit' or 'single-parent family' have been applied, although data were available to do so. Such indicators are often applied to 'measure' social exclusion. The reason for this is that the assumed relationship between unemployment

Figure 11.3 Relative deprivation in Amsterdam, 1st and 4th quartile, 1994

Source: Surveys AME, University of Amsterdam

and social exclusion still has to be proved. The unemployed, certainly in the Dutch context with high unemployment benefits, may be able to maintain and to develop social networks and to participate in all fields of urban life. Engbersen (1990) has shown that some categories of Dutch unemployed have very busy lives indeed. Consequently, we were keen to look for more direct indicators of social participation and for indicators of subjective feelings of loneliness.

Two indicators for social participation were used. Almost 60 per cent of the respondents appeared to stay at home for five or more evenings a week (almost always at home). Some 33 per cent of the respondents never went to the pub, cinema or theatre. Their participation score was modest. Most of the analyses generated identical conclusions using either of the indicators for social participation. For that reason we will predominantly focus on the presentation of empirical evidence by referring to the second indicator. For reasons of convenience we call that indicator social participation (moderate versus sufficient).

Segregation and social participation

People living in an area with a relatively large number of people who lack the prospect of upward social mobility appear to be socially excluded more often

Table 11.2 Association between segregation and social participation

Segregation (%without prospects within a radius of 500 m)	*Social participation*		
	Moderate (%)	*Sufficient (%)*	*Number*
< 4	26	74	1,348
4 – < 8	32	68	1,480
8 – < 12	39	61	781
≥ 12	46	54	382
Total	33	67	3,991

than people living in an area with relatively few others who lack upward social mobility prospects (Table 11.2).

The most logical explanation for this relationship is the labour market situation and the prospect of upward social mobility of individuals through this. We have already mentioned the higher probability of being socially isolated once someone is living in an area with many socially isolated people. Whereas 32 per cent of the respondents with prospects of social mobility could be labelled as socially isolated, 48 per cent of the respondents without prospects of social mobility were labelled as such. So there is a clear association between the prospects of social mobility and social participation at the individual level. However, surprisingly, the information provided in Table 11.3 shows us that the association between segregation and social participation persists for each of the categories of respondents (with or without prospects of upward social mobility themselves). In other words, high-skilled employed persons face a significant effect from the concentration of poverty on their own social participation, as do low-skilled and unemployed people. High concentrations are related to higher percentages of persons with only moderate social participation scores.

Table 11.3 Association between segregation and social participation, controlling for the respondent's individual prospects of upward social mobility

Segregation (% without prospects within a radius of 500 m)	*Prospects of social mobility*			
	with % moderate participation	*N*	*without % moderate participation*	*N*
< 4	26	331	34	20
4 – < 8	30	423	52	48
8– < 12	37	276	47	30
≥ 12	43	155	69	20
Total	31	1,110	48	118

Table 11.4 Percentage with moderate social participation (additive model)[a]

	%
Average	33
Segregation (% without prospects within a radius of 500 m)	
< 4	–3
4 – < 8	–1
8 – < 12	3
≥ 12	8
Education	
low or medium level	4
high level	–14
Age	
younger than 30 years	–21
30 years and older	7
Income per household, net per month	
< Dfl. 1,500	13
Dfl. 1,500 – < Dfl. 3,000	2
≥ Dfl. 3,000	–16
Ethnicity	
Mediterranean, Surinamese or Antillean	18
other	–2
Household type	
single person	–10
multiperson without children	8
multiperson with children	5

Note

a ANOTA-coefficients were calculated. Only the main effects of two-dimensional relations between all variables mentioned and the dependent variable 'social participation' are taken into account in these ANOTA models (cf. Deurloo 1987). The models applied allow for addition and subtraction of the coefficients to and from the overall probability score. If nothing else is known of the respondent, there is a chance of 33 per cent he or she has a moderate social participation score. But if we know he or she is living in an area with a relatively high percentage of inhabitants without prospects of social mobility, the chance will have risen to 41 per cent. If we know the person is older than 30 years the probability will rise to 48 per cent, etc.

To investigate the relation between segregation and social participation controlling for other relevant variables, we first looked at all direct relationships between social participation and the relevant variables (Table 11.4).

From this analysis it becomes clear that variables such as age and ethnicity, and also education itself (not linked to employment) and income, are all clearly related to people's level of social participation. Detailed analysis of the relationship between income and social participation reveals that there is no significant difference between low and medium incomes as far as social participation is concerned. Only higher-income groups show higher social participation scores.

In spite of the clear relationships between the variables mentioned and social participation, the link between segregation and social participation appears to

Table 11.5 Association between segregation and social participation (percentage with a moderate score), controlling for age

Segregation (% without prospects within a radius of 500 m)	*under 30 years of age*		*30 years and older*	
	% moderate	*N*	*% moderate*	*N*
< 4	7	26	33	325
4 – < 8	12	56	42	415
8 – < 12	21	52	47	254
≥ 12	17	19	58	156
Total	13	153	41	1,150

remain virtually unchanged if controlled for these variables. We calculated various interaction models, among them a series with education, age, segregation and social participation, in order to trace possible interaction effects. The interaction between segregation and age is the only one that appears to be relevant. For that reason the relationship between segregation, age and social participation is also shown in a cross-table (Table 11.5).

The association between segregation and social participation appears to exist for both age categories, but is somewhat stronger for older people.

Analyses in which other relevant dimensions are controlled for, such as ethnicity and income, provide identical conclusions. Within each of the categories distinguished, the association, central in this contribution, remains the same. And this relation holds for each of the combinations of individual prospects of social mobility and ethnicity as well: 47 per cent Turks, Moroccans, Surinamese or Antilleans who are low-skilled and unemployed are likely to be socially participating only moderately. However, if such persons live in an area in which at least 12 per cent of people lack social mobility prospects, the probability score rises to 71 per cent. For indigenous autochthonous Dutch who are low-skilled and unemployed there is a 47 per cent likelihood of only moderate social participation. For this group, the figure rises to 63 per cent in an area in which such problems are concentrated.

Conclusions

A preliminary and unexpected conclusion which can be drawn is that the segregation of poverty affects people's social participation (however defined) to an important degree. Apparently, only small differences in terms of segregation or concentration of poverty, as well as only low levels of segregation, are sufficient to generate substantial differences in terms of the percentage of people with reduced participation in society. Even in the Dutch situation, where only moderate levels of segregation are encountered and differences between neighbourhoods and between individuals are relatively small, such effects can clearly be found. These results are in line with a study by the Dutch Social and

Cultural Planning Agency, which concludes that ethnic segregation in primary schools in the Netherlands has an effect on success in education (Sociaal en Cultureel Planbureau 1996: chapter 6). However, unlike residential segregation, segregation in schools is considerable in the Netherlands.

These findings indicate that differences in social participation develop, even within the context of strongly redistributive welfare states. Even low levels of neighbourhood concentration of poverty have negative effects on a person's social opportunities. The high spatial concentrations which Massey and Denton (1993) refer to do not seem necessary to create substantial effects.

Two possible interpretations may lie behind the findings presented here. The first is that social processes such as the *stigmatisation* of a neighbourhood, with all the negative consequences likely to be associated with that (being avoided or excluded by certain employers, for instance), will produce negative effects very quickly, and are dependent not on absolute levels of deprivation, but on relative levels. However, the consequences can be just as disastrous. A second, somewhat related interpretation, is that even small concentrations of poverty produce sufficient *negative examples* for others (particularly children), in terms of role models, to raise the probability that they start to behave differently from others in society: for example, social pressure may create an atmosphere in which higher education is not aimed at, or voting is not engaged in.

What seems to be evident, however, is that the planned reformulation of the welfare state, which is currently going on in virtually all areas of society, from taxation to health, from incomes to housing, from employment to benefits, will have major effects, increasing the segregation of the population as well as increasing the proportion of the population that loses contact with society. The only positive development could be an increase in participation in the labour market. Unfortunately, the mechanisms described give no room for other interpretations.

However, at this point we have to stress that the research findings are based on a relatively thin empirical base. The subject is very important, however; even more so in the light of the restructuring of the welfare state. Future research, some of which is well under way already, may provide additional answers to the question dealt with here. Furthermore, it is of great importance to continue international comparative research, to investigate the relative influence of different welfare state models, and different levels of spatial segregation in cities, on the probability of becoming socially excluded. That is one of the challenges in urban research in the years to come.

Note

1 Analyses were performed using various radii. The chosen radius appeared to be best in terms of detail and in terms of differentiation.

12

THE DIVIDED CITY?

Socio-economic changes in Stockholm metropolitan area, 1970–94

Lars-Erik Borgegård, Eva Andersson and Susanne Hjort

Increasing socio-economic polarisation – an international phenomenon

For many years welfare indicators such as those relating to housing, labour and salary conditions improved in most countries in the western economy, as did the general standard of living, including health. However, recently there have been signs of a dismantling of welfare systems. This restructuring of the world economy has been described and analysed by several scholars over the last few years, with themes such as global shift, divided cities and dual cities. A sign of increasing polarisation is the international discussion of the so-called two-thirds society (Olsson-Hort 1992), meaning that most of the population has resources while a minority lacks them. A growing international literature points in the same direction and has brought attention to polarisation, marginalisation, poverty and lately to the problem of social exclusion (*Urban Studies* 1994, *Built Environment* 1994, Scott 1994, Funken and Cooper 1995).

Income distribution, housing conditions and spatial location are linked together. Households with low incomes or those unemployed have had a tendency to concentrate in specific locations during most of the industrial era. The unemployed have been found in poor housing, traditionally in inner-city areas. The connection between unemployment and housing problems is not only a problem in countries with a large private housing market, but also in countries with plenty of social housing.

Recently these problems have emerged especially in large housing estates, but also on the periphery of towns and cities. Unemployment is only one important dimension showing changes in individual lives caused by changes in the macro-economic structure (McGregor and McConnachie 1995). The basis of the ongoing debate on social exclusion is to be found in the changes in the macro-economic structure. In these discussions is also included the belief that the problem of spatial concentration of the disadvantaged in society has grown over time. The crucial point is the fact that disadvantaged people tend to cluster, voluntarily or involuntarily, in isolation from mainstream social and economic activities. This means that a growing number of households find considerable restrictions on access to services, facilities and networks in the wider economy and society, that is, they are socially excluded (McGregor and McConnachie 1995). There is debate over what the reasons for this are and how to cope with them, but there are no self-evident or easy solutions (Power 1996). Programmes have been started, concluded and evaluated, but it may be that the root of the problem is beyond these trials executed at a low geographical level. Here we find suggestions that the real source of the problem is to be found at higher, global and national levels (Sassen 1994a) and that the measures used are either too weak or not sufficiently multi-sectoral.

The basic propositions in Sassen's 'global city thesis' (1994a) hold that the restructuring of the economies of dominant cities in the global urban hierarchy, such as New York and Los Angeles, causes social polarisation. This global city thesis has been elaborated in earlier chapters. The claim that developments in global cities are paradigmatic for those in other cities is important.

The theory presented by Sassen, suggesting a similar polarisation in all global cities due to economic changes, is rejected by Hamnett (1994a). He pinpoints evidence from the Randstad region and argues that there has been a certain professionalisation of the labour force but no real polarisation, at least not during the 1980s. In a later paper, Hamnett (1996) further elaborates his criticism of Sassen's theory. He claims that social polarisation of the occupational structure and incomes in global cities cannot be viewed as inevitable or as a direct result of economic restructuring.

The two different standpoints of Sassen and Hamnett are elaborated in a more moderate way by Murie and Musterd (1996), who argue that the explanation for polarisation lies somewhere between the two views presented above. The ideas about a global economy and global influences on both market systems and on public policy draw a picture of convergence in processes and policy, and, most importantly, they offer expectations in terms of outcome with similar patterns of polarisation and division. Murie and Musterd argue for a more differentiated picture of the current and expected outcomes of residential segregation and polarisation. They find that the emerging patterns of social segregation in the Netherlands and in Britain are very different. It is therefore of great importance to relate globalisation and its influences to other factors such as variations in housing finance and policy, the degree of social and income inequality, and the interventions of the welfare state (Murie and Musterd 1996). Globalisation, on the one hand, and the specific context within a country on the other, create a special residential pattern concerning social polarisation and residential segregation. In a recent paper Hjarnø (1996) argues that Copenhagen doesn't fit into the polarisation thesis, because the social welfare state gives support to those in need and does not allow salaries to polarise too much.

This chapter will concentrate on the Swedish experience. We start with a section in which the problem addressed in the present study is examined, together with a discussion on Swedish welfare and residential segregation focused on income distribution. The next section elaborates ideas about the restructuring of the Swedish economy, followed by a presentation of the data relevant to the area of the study. Results of the study are followed by a short discussion, summary and concluding remarks.

Problem, aim and questions

Income equality has long been an overriding political goal in Sweden. Increasing improvements in income equality took place in parallel with the construction of the welfare system in the Scandinavian countries. In Esping-Andersen's (1990) classification of different welfare regimes, Sweden stands as the model for the social democratic welfare state.

Besides an even income distribution between households, the government has also aimed at mixing different groups of households in 'integrated housing', which ideally is a mix of households with different demographic, socio-economic and ethnic characteristics. However, there are many forces working in an opposite direction, that is, towards a separation of households with different incomes and socio-economic characteristics. The problem discussed in this chapter has to do with whether growing inequalities in household income and socio-economic division are discernible in the spatial pattern and, if so, at what geographical level such a development can be seen. The aim of this chapter is to analyse changes in the patterns of socio-economic polarisation, measured by mean income, in Stockholm metropolitan area during the period 1970–94.

The questions at issue are as follows.

1. In what way have incomes developed during the period 1970–94? Has divergence or convergence occurred at the municipal and housing area level? Are there differences between different geographical levels?
2. Is there a spatial pattern of income distribution between municipalities, between inner and outer areas of the city, and between the traditionally rich northern region and the poorer southern part?
3. What are the possible explanations for the development of convergence or polarisation and for the spatial distribution of incomes?

The hypothesis is that the differences in income between municipalities have decreased over time while differences between housing areas have increased; in other words, the increasing economic polarisation can also be measured in spatial terms.

Polarisation could be measured in many ways, such as with respect to class structure and economic structure. In this study the data are limited to mean incomes, giving one indication of the most important welfare variables. However, there is a strong correlation between income and socio-economic class (determined by occupation, education and income). Also there are strong negative correlations between income and tenure status and between income and recent immigrant households (SOU 1990a: 20, Statistics Sweden 1995).

The Swedish discussion on welfare and residential segregation

In this survey of literature two lines of discussion are followed: studies of the standard of living and studies of segregation, especially on the basis of income distribution.

Studies of standards of living – economic resources

Swedish welfare policy has been very much focused on economic resources. One prominent part of the so-called 'Swedish welfare model' has been to look after marginal groups in the welfare system (Johansson 1970). However, spatial analyses have not been extensive and the connections to housing have not been evident. Analysis of socio-economic class has been most prominent. A decade-long inquiry, *Inequality in Sweden* (Vogel *et al.* 1987), shows changes in living conditions between socio-economic household groups as well as between demographic groups. In addition, Halleröd (1996) has shown the increasing polarisation between household groups according to level of education.

In spatial terms a couple of studies have been carried out describing regional income distribution. Mean income varied mostly between regions in the 1970s. Regions with high incomes were the metropolitan areas, and those in the periphery had the lowest mean incomes (SOU 1974: 10, appendix). One explanation for this pattern was provided by differences in the labour participation of women. Other explanations were differences in occupational structure and level of salaries and regional variation in an index of consumption. Since the beginning of the 1970s, labour participation has levelled out regionally, and there has been a homogenisation of salary structure and occupation structure (Work and Leisure 1993). Using mean values, a general finding is that a regional convergence of incomes has occurred over time.

Studies of segregation measuring income

Most Swedish studies on housing segregation have focused on socio-economic variables, especially class structure and ethnicity. In Danermark's (1983) study of segregation in three regional capitals in Sweden – Örebro, Norrköping and Karlstad – he calculated the spatial income distribution as well as class segregation. He found that the wider the gap between income classes, the further apart the groups lived. High-income and low-income groups lived most isolated from each other. The middle-income earners were the most evenly distributed households, with low segregation indices. Comparing the variables class and income, Danermark pointed to income as the more distinctive segregating variable.

In the report on welfare and segregation (SOU 1990b: 20), segregation is seen as resulting from differences in income, connections and knowledge. Socio-economic structure in different areas changed in the whole country but particularly in the metropolitan areas and mostly in the Stockholm region, from 1980 to 1985. The redevelopment of the inner city led to a replacement of people with few resources by people with more resources. The proportion of unskilled labour in owner-occupied housing increased somewhat during the period, while the highly skilled remained heavily over-represented in these areas. The so-called Million programme areas, which were built between 1965 and 1974 in the suburbs, experienced a high rate of mobility and a high level of so-called problem households and immigrants. This was the state of the metropolitan areas during the first part of the 1980s. During the latter part of the 1980s, the patterns of segregation have, according to many reviewers (SOU 1993: 91), been reinforced. Biterman (1994a, 1994b) accounts for dissimilarities between municipalities in the Stockholm region during 1970–90, but there is no statistical evidence for an increasing polarisation. However, looking at the Million programme areas, Biterman finds an increasing number of low-income earners, immigrants, single mothers and low-skilled households, indicating that a marginalisation process is under way.

A lot of concern has been focused on the 34 Million programme areas that were targeted in the Stockholm region. These areas are predominantly housing blue-collar households, single parents with children, young and old single persons and immigrants, especially those who have come to Sweden recently. The mean population of these areas is 6,000 inhabitants, with the largest having 15,000 and the smallest 2,000. The population of the Million programme areas in the Stockholm region constitutes approximately 15 per cent of the regional total. In a recent follow-up to Biterman's (1994b) study, the areas were classified according to the ratio between low/high-income earners during the period 1975–93 (Inregia 1997). The main results showed increasing poverty in the 34 areas for the period 1975–85, a modest recovery between 1985 and 1990 and a regression to increasing poverty between 1990 and 1993. In general terms the social indicators, measured by households on social welfare and rates of unemployment, increased in the entire region in the 1990s. However, the increase was greater in the Million programme areas. Therefore one conclusion is that an increasing social and economic polarisation has been taking place between residential neighbourhoods, especially in the 1990s.

In a cartographic study of small geographical areas in the Stockholm region, 1970–90, the most discriminating variables between residential areas are education and income (Regionplane- och statistikkontoret 1995). A comparison between variance in incomes for all municipalities in the Stockholm region and 110 statistical areas within the Stockholm municipality in 1994 shows, not unexpectedly, a bigger variation at the smaller spatial scale (Bernow and Strömqvist 1995). Thus the above studies indicate an increasing economic polarisation during the 1990s in the Stockholm region, especially at the lower geographical level.

Concepts of restructuring the economy – the Swedish case

Processes of social polarisation, as Murie and Musterd (1996) point out, need to be seen in a broader context, where consideration is given to variations in national policies in different welfare areas. The driving forces behind socio-economic polarisation are found on structural – international and national – local and individual levels and these different levels are intertwined. The process is dynamic and the driving forces can be strengthened or weakened over time. Changes on the different levels will have a spatial impact on both the distribution of population and on the distribution of its socio-economic characteristics.

On the international level there have been many changes taking place during recent decades. As mentioned before, globalisation and restructuring of the economy are two elements emphasised by many scholars (e.g. Sassen 1994a). These global changes have affected Sweden in that the traditional manufacturing sector has decreased and the numbers of people involved in the public sector, private enterprise and education and information technology have increased (Axelsson and Fournier 1993). Another change on the international scene is the increase in the number of immigrants that has taken place over the last decades, and, more recently, the increasing number of refugees (Castles and Miller 1993). At the level of national economies a social polarisation can also be seen in the increasing differentiation between high-skilled and low-skilled individuals.

Sweden – national level

Sweden has a long tradition as a welfare state. In the welfare regimes elaborated by Esping-Andersen (1990) Sweden is classified as the basic model of a social democratic welfare state, which means a strong public service sector distributing welfare facilities to all households having needs such as work, housing, day care, elderly care and so forth. The service sector, including a large public sector, comprises 60–70 per cent of the Gross National Product.

One of the ideological cornerstones in the Swedish welfare state is the idea of equality between different households despite demographic, socio-economic and ethnic characteristics as well as where they live. The idea of 'integration', the mixing of households with different backgrounds in housing areas, can be seen as a spatial equality goal. To attempt to achieve these goals, the state has transferred resources to households: pensions, social welfare and so forth. It is obvious that the equality goal worked according to plan until the beginning of the 1980s. Measured by the GINI-coefficient, the figures showed a decrease in inequality between households until 1981, and then an increase, which accelerated in the mid-1990s. Even if Sweden, by international standards, has a relatively even income distribution (Ginsburg 1992), the tendency towards inequality is clear (Statistics Sweden 1996).

In general terms the economy was strong in the 1960s, and in the 1970s the restructuring of the economy was evident. Many formerly strong sectors were almost wiped out, but two major factors helped alleviate the negative consequences. One was building up the public service sector, which was spatially decentralised, and the other was increasing the number of women recruited to that sector. The period 1980–90 displays two different economic phases: the slow economy at the beginning of the 1980s, followed by an increase in GNP and a booming economy at the end of the decade, peaking in 1990. During the early 1990s the unemployment rate increased substantially, from its lowest point, 2 per cent in 1989, to 8–12 per cent two years later (depending on what groups are included).

Recent changes in the labour market have caused problems for individuals, households and local authorities as well as for the general social welfare system. Until recently the flows on the labour market were balanced. An inflow of young people was taken care of in the employment sector, and increasing unemployment rates during one or two years were later met by an increase in employment rates. The early retirement rate was modest. However, since 1990 a new situation has arisen. Figure 12.1 indicates the unbalanced situation, where the starters, composed of young people, together with recently arrived immigrants and refugees, want to join the labour market, but are met by

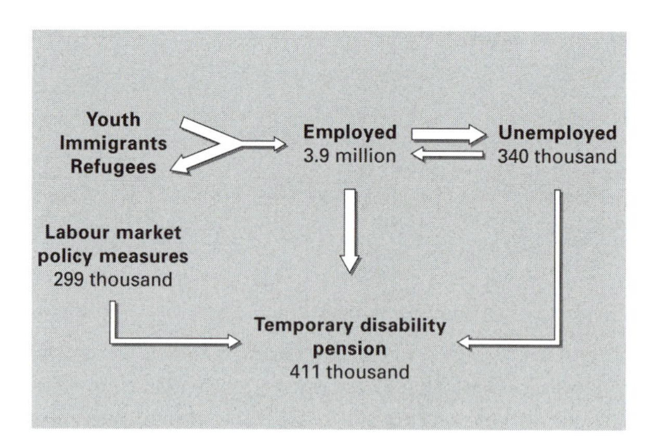

Figure 12.1 Recent changes in the labour market sector (1994)

Sources: Social rapport 1994, Statistics Sweden 1996

numerous barriers, indicated by the reversed arrow. There is also a net outflow from employment to unemployment, which is also the case from employment to early retirement. Inflow to employment is weak and outflow is substantial. Recent studies show a big workload of overtime in the employment sector.

Some recent changes in the social welfare system have been cuts in unemployment subsidies, in housing subsidies, in 'sick days', and so forth. These cuts result in increasing numbers of homeless people, evictions and those on social security (Social rapport 1994, Sahlin 1996).

Stockholm – local level

Structural economic changes in Sweden and particularly in the Stockholm region during the investigated period are important components in the explanation for the development of social polarisation. The Metropolitan Commission has demonstrated the economic restructuring within which a professionalisation of Stockholm has taken place (SOU 1990b: 36). Well-educated and high-income earners migrated to Stockholm, where there were greater opportunities to get a high-status job than in the rest of the country. Andersson (1996) uses the idea of Stockholm as an 'escalator region', where young people especially can take another step in their professional careers. Stockholm inner city experienced dramatic demographic and socio-economic changes during the 1980s, with a new class of top professionals 'invading' the 'old widow areas' (Borgegård and Murdie 1994). Stockholm has also strengthened its role as the prime city of Sweden, with a high concentration of main offices, international banks, insurance companies and high-tech corporations (Borgegård and Murdie 1994).

The housing market has undergone dramatic changes in the Stockholm region during the last decades. There are two main elements to be looked at – the shortage of housing in the expanding Stockholm region and the equality element, 'good housing for all households'. In the Million housing programme, there was concentration on public housing, especially in the outer southern suburbs, in order that a better balance of population in Stockholm could be achieved. A new rent system was introduced, equalising rents for housing of equal standard, regardless of location. In the 1960s the former city centre of Stockholm city was demolished, rebuilt and completely modernised, and the former large immigrant population in this reception area moved to the suburbs, indeed to the same areas as many newcomers from the rest of Sweden. After the Million housing programme period, the government increased the production of single-family housing, giving rise to a massive exodus of households with children to these areas. The loans were favourable and interest rates deductible on income taxes. Heinstedt (1992) has demonstrated the big change in the public housing stock of Stockholm county 1960–90, based on five-year census data, showing an increasing over-representation in the public housing stock of socio-economically disadvantaged households, of foreign-born households and of households on social welfare.

Stockholm – people and housing

Structural economic changes in Sweden and particularly in the Stockholm region during the period under investigation are an important part of the explanation for the development of residential segregation. During the last few years Stockholm has experienced many changes. These changes can largely be said to follow the economic fluctuations, but with some delay. Also, the earlier mentioned structural changes towards a professional society, seen in most post-industrial societies, have led to an increase in offices and shops and a decrease in housing in Stockholm's centre. The housing that is left is expensive and inaccessible, compared to the Million programme areas in the outer suburbs.

With the slow economic growth that followed the oil crisis in 1973, costs for social transfers and public expenditure increased and the view of the welfare state changed in public debate. Social expenditure was seen as the largest problem that the government had to deal with. Despite this it is hard to find concrete evidence of an obvious trend break in the 1980s (Fritzell and Lundberg 1994). In the 1970s and 1980s there was exceptional prosperity in the housing market. It was comparatively easy to borrow money from the banks to finance housing. Income development at the beginning of the 1980s was characterised by a downturn in purchasing power, the disposable income per person sank, but during the late 1980s it rose again. From a segregation perspective the differentials in income are more interesting, and they increased during the entire 1980s, when there was also a boom in the economy. The old 'rural' Stockholm was finally pulled down and replaced by modern buildings (Olsson-Hort 1992).

Since the first few years of the 1980s Stockholm county has had positive population growth. The only decade in the twentieth century with a population decrease was the 1970s (Bernow and Strömqvist 1995). Today there are 1.7 million people living in Stockholm county. In the 1980s, during the economic boom, the population increased by 65,000 people, a growth of 4 per cent. The shortage of housing in the region was and still is acute. Despite this, there was very little construction of dwellings during the 1980s. Large redevelopment programmes have been carried out, but they have, in fact, led to a reduction in the numbers of available housing units, because small flats were made into larger ones. The shortage of housing affects various groups differently, which is one reason for residential segregation. It was not until the 1990s that political changes started to show, for example privatisation of the housing market. Building in the Stockholm region has sunk during the 1990s to the lowest levels since the beginning of the century. Still, there is considerable population growth in the region (Bernow and Strömqvist 1995), resulting in an increased density of dwellings.

The choice of units of investigation – municipalities and neighbourhood areas

The public sector is very large in Sweden compared to other countries; Sweden also has very high levels of taxation. These two phenomena are connected. One thing that differentiates Sweden from many other welfare states is that most taxes are paid to local government. The municipalities have the right to tax all taxable objects in Sweden. The municipal level of taxation lies between 30 and 34 per cent irrespective of income, while the state income tax is progressive. This is the reason for devoting special attention to income distribution at the municipal level. It means that public service supply can vary between municipalities and, with that, people's living conditions.

The municipalities practise different types of housing policies, for example in relation to public housing. This varies considerably between municipalities and it has been demonstrated that there is a selection process taking place which is splitting households, predominantly due to economic resources, into the public and the privately owned single-family housing sectors. This polarising process has been going on for the last few decades. The question arises as to what aggregated spatial level the dividing line between different groups of households lies on, the municipal level or at the neighbourhood area level.

Study areas

In Stockholm county the population is 1.7 million people, divided into 25 municipalities. Stockholm municipality in the centre of the region accounts for almost half the population (approximately 750,000 inhabitants). In this study two municipalities on the periphery, Norrtälje in the north and Nynäshamn on the south, are excluded, leaving 23 municipalities to be accounted for.

At the lowest geographical level 330 local areas in the Stockholm region have been used. The average population size in these areas varies between 4,000 and 10,000 people. The division is based on socially meaningful units of area such as neighbourhood units (Biterman 1994a). The measure of income

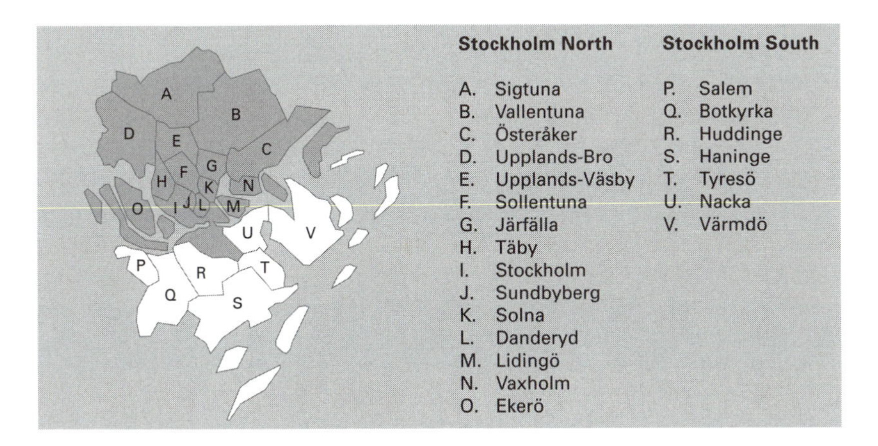

Figure 12.2 Stockholm region by municipalities

used in this study is gross income, which differs from disposable income. The latter includes transfer payments from the government to the household. The period studied is 1970–94. A problem of measurement concerns time. The years that have been chosen for measurement are 1970, 1980, 1990 and 1994. During this period the consumer price index has increased substantially. Still, the absolute income is not of interest in this study, but rather the variation between and within municipalities at different times. Therefore it has not been necessary to adjust the income values.

The geographical limits are the municipalities of Stockholm county. In this study we have used the Stockholm region plus the municipality of Södertälje. This means that all the municipalities in Stockholm county are included except for Norrtälje and Nynäshamn.

Empirical results

Income distribution – municipal level

In general terms Swedish income distribution became more even during the 1960s and 1970s with the lowest figure of the GINI-coefficient 0.203. After that the figure has risen gradually, and in the 1990s quite substantially. In 1994 the coefficient reached a value of 0.288 (Figure 12.3). The question is whether the more uneven income distribution has a spatial impact – in this case with reference to the Stockholm region.

From the statistical analysis at the municipal level of Stockholm region two significant points can be made. First, there is a convergence of incomes, as measured by mean incomes, between the 23 municipalities. In 1970 the standard deviation compared to mean income value accounted for 13 per cent of the mean

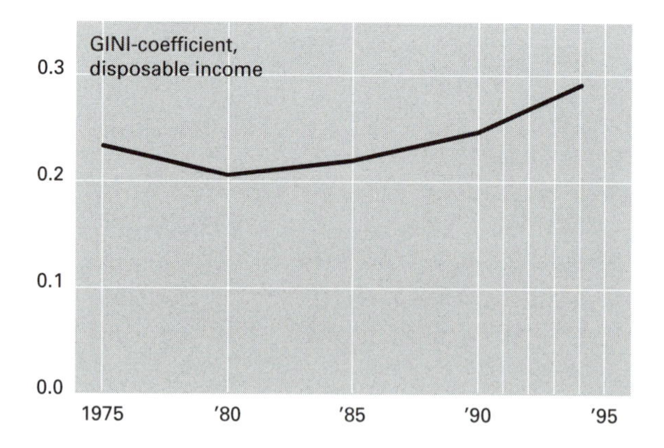

Figure 12.3 GINI-coefficient, 1975–94

Source: Statistics Sweden 1995

value, and this figure decreased gradually to 6 per cent in 1990. The over/under-representation of mean incomes, measured by the ratio of individual municipalities to the mean value of all municipalities, also converged over time. The highest mean income value of over-representation was, for 1970, 1.45 for the 'high status municipality' of Danderyd. The ratio decreased to 1.17 in 1990, giving evidence for convergence of mean incomes on the municipality level.

The ratio between the upper and lower quartiles of income was measured as an indicator of equality at the municipality level. Income distribution by income classes was calculated. The ratio between the top and lower quartile is quite modest (1.7–3.7), for all observations in 1970, 1975, 1980, 1990, 1993 and all municipalities. The maximum ratio between the highest mean income quartile and the lowest is almost 4:1 and the minimum ratio is almost 2:1. Over time convergence of income could be confirmed for all municipalities and most clearly in the rich municipalities of Danderyd and Vallentuna.

During the period 1980–93 higher mean incomes are seen in all municipalities but also a decreasing gap between high- and low-income municipalities. However, in the light of recent GINI-coefficient increases in the 1990s, it could be hypothesised that income changes were emerging between 1990 and 1993 after the new tax regulation in 1991 (Ringqvist 1996). In order to measure the relative change of mean income at the municipality level, the ratios between 1980/90 and 1990/93 have been calculated (Figure 12.4).

During the period 1980–90 the pattern shows a relative increase in mean income for the outer suburbs. Recent changes in 1990–3 show a different picture, supporting the increasing polarisation hypothesis. Environmentally attractive areas in the countryside and areas close to water have increased their population numbers by in-migration. These municipalities are deliberately attracting wealthy households.

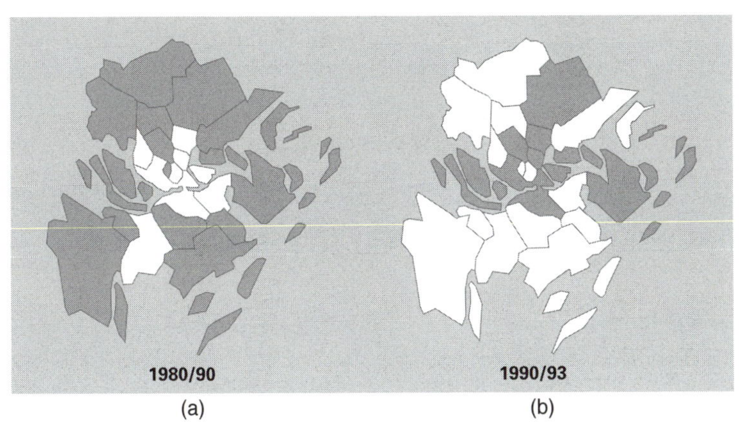

Figure 12.4 Changes in mean income ratios at the municipality level 1980/90 (a) and 1990/93 (b) in the Stockholm region

Source: Statistics Sweden, income statistics

Table 12.1 Mean income ratios between areas of Stockholm, 1970–93

Geographical area	*1970*	*1980*	*1990*	*1993*
North/south	1.078	1.067	1.055	1.095
Outer north/outer south	1.106	1.097	1.085	1.111
Inner north/inner south	1.058	1.056	1.047	1.086
Outer north/inner north	1.007	1.043	1.070	1.014
Outer south/inner south	0.963	1.004	1.033	0.991

Source: Statistics Sweden, income statistics

Regional level

Mean income values have been used to calculate ratios between inner city and outer suburbs, and the north and south over the period 1970–93 (Table 12.1). A ratio of 1.078 means a difference in mean income of almost 8 per cent in favour of the 'north'. Even if the differences seem to be quite small, there are some general trends to be seen from the figures. The north/south ratio was declining at the regional level between 1970 and 1990, but recently the gap has widened again. One reason for the increasing gap between north and south lies in the attractive areas built out to the north and a relatively large increase in high-income-earner salaries after the tax reform of 1991.

This may also be the explanation for the recent widening gap between outer north/outer south and inner north/inner south. In the case of outer north/inner north an explanation could be the villa boom of the 1970s, when households moved out of the inner city and settled in the northern suburbs. The levelling out of the differences could be seen in the light of an increasing gentrification of the inner city of Stockholm. The same reasons lie behind the ratio of outer south/inner south. In general terms the changes could be summarised as decreasing differences in income values during the period 1970–90 and recently increasing gaps between regions.

Neighbourhood data

Analysis of detailed maps of the 330 residential areas in 1980 and 1990 shows a very dispersed picture. Four main patterns can be detected, though. First the inner city of Stockholm municipality indicates low income values, due to demographic and socio-economic structures. Old people are over-represented in inner Stockholm (Borgegård and Murdie 1994). Exceptions to low-income values are found in the old multi-family brick building area of Östermalm (Stockholm inner city, east), with high-income households. Traditionally this is the wealthy part of inner-city Stockholm (William-Olsson 1961). Second, in a circle approximately 5–7 km from the inner city there are wealthy areas, formed of single-family housing areas. Many of these areas were built in the 1960s and 1970s and are now housing people with high education, good jobs and relatively low housing costs.

Many of these wealthy areas are close to poor areas, with predominantly multi-family public housing areas built during the Million programme. Third, the north–south divide is obvious, but not dramatic. There are poor areas in the north and rich areas in the south, even if there is an over-representation of a richer north and a poorer south. Finally, the pattern has been stable between 1980 and 1990, which is expected. No dramatic changes have been seen. A strong component of stability is the tenure division and the strong correlation between tenure status and income and socio-economic status.

Recent data from the Statistical Planning Office in Stockholm City provided income data for 1992 and 1994 for the 119 statistical areas of the city (DN 1996). Even though the time perspective is very limited, the changes are quite dramatic. Incomes are measured for households of two adults and two children, one of the common classes for calculating income data. Mean income increased by 7 per cent between 1992 and 1994. The correlation between incomes in the two years is high ($r = 0.94$), which is expected. More interesting, however, is that the changes were squeezed, which is demonstrated in Table 12.2.

The mean incomes of the 119 *statistical areas* were compared. The value of 2.23 has been arrived at by dividing the maximum value of mean income by the mean income of all statistical areas. The over-representation of incomes in high-income areas increased over time and the under-representation in areas with low incomes also increased. An increasing income polarisation between households and residential areas could be seen between 1992 and 1994.

Another set of income data was calculated comparing the mean incomes in 1990–3 for the 110 *parishes* in the Stockholm region. The results showing the ratio between incomes in 1993 and those in 1990 indicate an increasing gap between those parishes with high mean income earners and those with low mean income earners. Most striking was the increase in ratio between 1993 and 1990 for parishes with high mean income earners.

Combined with data from the Stockholm Office of Integration (1996), these data allow some statistical explanations for the polarisation process to be given. There are strong negative correlations between high unemployment rates, concentration of immigrants and low-education households. These background factors are strongly negatively correlated with incomes. There is also a strong spatial component, giving support for the existence of a polarisation of the city

Table 12.2 Over- and under-representation of incomes in Stockholm city statistical areas, measured by ratio, 1992 and 1994

	% of population						
Year	*Min*	*10*	*25*	*50*	*75*	*90*	*Max*
1992	0.64	0.84	0.86	0.96	1.10	1.22	1.77
1994	0.57	0.77	0.82	0.95	1.10	1.33	2.23

Source: DN 1996

between areas with resources, especially the north-east (Östermalm), and those with scarce resources (Rinkeby-Tensta, Skärholmen) in the north and south. The pattern is still not easy to decode, because there are areas within the 'poor' with wealthy people and poor households close to wealthy areas. An explanation for this mosaic pattern is provided by the tenure structure, which gives a general clear distinction between an over-representation of richer households in owner-occupied areas and an over-representation of poorer households in municipal housing areas (Borgegård and Murdie 1994).

Summary and conclusion

Sweden has long been regarded as the model of the welfare state, which was true for the 1970s and 1980s. However, many external changes have had an impact on the living conditions of the households. In this chapter, three main questions have been answered.

1. *In what way have incomes developed during the period 1970–94 towards divergence or convergence at the municipal and neighbourhood area level? Are there differences between different geographical levels?*
   One main component in the Swedish welfare state ideology is equality between individuals and households. Measured by the GINI-coefficient, income distribution decreased until the beginning of the 1980s, since when it has been increasing continuously, especially in the 1990s. In spatial terms, measured by mean income on the municipality level for the 23 units in the Stockholm region in 1970, 1980, 1990 and 1993, a process of convergence is seen, at least until 1990–3, when a slight change towards increasing mean incomes emerges. During the period 1980–90 an increasing income gap is seen at the low geographical scale, measured by the 330 neighbourhood areas. This polarisation has increased during the 1990s. The hypothesis that differences in income between municipalities have decreased over time (1970–90), while differences between housing areas have increased, has thus been confirmed. However, signs of increasing differences even at the municipality level have been demonstrated during the 1990s.
2. *Is there a spatial pattern in income distribution between municipalities, between inner and outer areas of the city, and between the traditionally rich northern region and the poorer southern part?*
   The spatial pattern of the traditional 'rich north' and 'poor south' is levelling out, measured on municipality and parish level over time. Recent changes in the 1990s do, however, indicate the emergence of an old pattern of a slight polarisation between 'north' and 'south'.
3. *What are the possible explanations for the development of convergence or polarisation and the spatial distribution?*
   Some basic changes have taken place in the modern welfare society, which can be exemplified by Sweden and specifically the case of Stockholm. A

> main component of the explanation for the polarised society is the restructuring of the economy, with a decrease in employment in the industrial sector and increase in the service sector. Changes in the economy were maintained by a successful, active labour market policy (Ministry of Labour 1988). However, there were long-term factors undermining the foundation of a functioning labour market. For the past two decades, until recently, the balance between people consuming and producing goods and services was clear. A short period of life was devoted to education and after that period jobs were available and retirement took place at 67 and more recently at 65. During the last decades the youth and education period has been prolonged while the retirement age is still 65 formally, but 58 statistically (Work and Leisure 1993). The ratio between individuals in production/consumption life-cycle phases has changed dramatically, which also means a reduction in those depending on earnings by employment. When the production base was founded on the industrial sector (employing 30–40 per cent of the population) and when the service sector took care of those entering the labour market and those unemployed from the industrial sector, the system was working, because the whole transfer system was based on taxation and high employment rates. The social welfare system was also dependent on inflation and loans from abroad. These changes are part of the explanation for the deteriorating welfare state of Sweden.

Sweden shows an increasing polarisation between households and between residential areas. However, put in perspective, it is still a country with a very egalitarian distribution of incomes and with a high standard of living. In the introduction to this chapter the question was raised as to how Sweden relates to the global polarisation process. Sweden is infected by and suffering from high unemployment rates. It has not adopted a policy of reducing salaries in the service sector in order to produce new jobs for those excluded from employment. The labour unions are still strong and still resistant to demands for low salaries for unqualified service jobs (LO 1996). A recent lively discussion in the autumn of 1996 and onwards on the social welfare system still indicates powerful support for medical treatment, elderly care and a strong school system.

With reference to the discussion by Sassen (1994a), Hamnett (1994a, 1996) and Murie and Musterd (1996), our results show that Sweden and the Stockholm region come closest to the more modest interpretation of the global city hypothesis. Sweden is closer to the western welfare states, with substantial government direction, which is also similar to the Danish experience shown by Hjarnø (1996).

The traditional Swedish welfare system is being restructured. Gradually the strong state interventions have been withdrawn in favour of local government. In line with changes in power and responsibility many of the regulations on the housing market have been made voluntary instead of compulsory. State subsidies to the housing market will change from a deficit to a net income gain for the state in a year or two (SOU 1996: 156). The formerly regulated rent-setting system will be gradually abolished. These changes are all signs leading to a more market-oriented housing market.

What the long-term consequences will be in terms of economic, spatial and cultural polarisation is unclear, but our results during the first half of the 1990s indicate a growing polarisation between households at a residential area level. However, further research should be focused on small residential units and on longitudinal studies of household groups in order to analyse long-term changes in welfare indicators. Emphasis should also be placed on the interplay between those households and neighbourhood units with particular negative social and economic indicators. International comparisons should be made to enable recent changes in the Swedish welfare state, and the consequences of these changes in terms of segregation and polarisation, to be placed in wider perspective.

13

(DE)SEGREGATION AND (DIS)INTEGRATION IN SOUTH AFRICAN METROPOLISES

Anthony J. Christopher

Introduction

The term 'apartheid' comes from the Afrikaans language and originally could be translated as 'separateness'. However, by the 1940s it had acquired additional connotations and become synonymous with the National Party of South Africa's political programme aimed at legislated racial separation. This programme was subsequently put into devastating effect. As a result the word has been absorbed into other languages to mean the ultimate form of ruthlessly applied and institutionally enforced racial segregation. Racially based separation was intended to operate at all levels of society. Thus, whether it was sitting on a park bench, standing in a queue in a shop or selecting a place of residence, there was to be no mixing of the White population with anyone else. Apartheid policies were enshrined in the law after 1948, to the condemnation of world opinion.

The legislated segregation imposed upon the population of South African cities has been lifted, but adjustment to the new freedom of choice following decades of rigorously applied apartheid planning is only just beginning. The impact of such a long period of institutionalised and all-pervasive practices is not readily erased either in people's perceptions or in reality on the ground.

Racially based segregation in South African cities before apartheid was akin to that encountered in other colonial cities of the period (Christopher 1992). The apartheid era, after 1948, resulted in the imposition of a far more ruthless and systematic approach to the issue, linked to doctrines of racial superiority. People were classified into discrete racial and ethnic groups, membership of which determined most aspects of their lives. Most significantly they were assigned legally defined areas within which they were required to live. The result was the emergence of the apartheid city with its extremely high levels of segregation (Christopher 1994a, Lemon 1992, Smith 1992, Swilling,

Humphries and Shubane 1991). A minor degree of relaxation of the very oppressive laws governing segregation was apparent between 1982 and 1990. Finally, the formidable array of laws, rules and regulations which enforced apartheid was repealed in mid-1991. Such was the thoroughness of the planning process and the ruthlessness with which the programme was executed that dismantling the apartheid city will be a long and difficult process, but clearly linked to the national Reconstruction and Development Programme outlined by the African National Congress (1994).

The development of the apartheid city

The creation of the apartheid city was a long and complex process with origins in the colonial period. The philosophy behind legalised segregation was the desire by the White (European) electorate to maintain and extend a combination of social exclusion and economic and political dominance over other groups in South African society. The social incompatibility of the various population groups was held to be real and permanent by government ministers and officials, while exclusion from areas of economic opportunity within the city was a strong motive in the drawing of both metaphorical and physical lines between the various groups. Segregation was officially viewed as a means whereby political and economic power could stay firmly in the hands of the White minority, which at the time constituted approximately a fifth of the total population of the country, yet controlled the political process.

Residential segregation had been practised in the colonial period before the establishment of the South African state in 1910. This had ranged from the construction of separate formal housing estates (locations) for the indigenous inhabitants to the erection of single men's barracks (compounds) for African transitory migrant workers. In addition in the Transvaal separate 'bazaars' for Indian settlers had been set aside adjacent to the White-dominated towns. Within the private land market racial exclusion clauses had been introduced in land transaction deeds to specify who might or might not own or occupy property in certain suburbs. The result was a wide range of segregation practices across the country and consequent highly variable levels of segregation (Christopher 1988).

After the formation of the Union of South Africa in 1910 White political pressures intensified in Parliament to make segregation practices more rigorous and uniform across the country. A barrage of legislation ensued. The provision of state funding for municipal housing under the Housing Act of 1920 was linked to the construction of segregated estates. Furthermore, all municipalities were required to provide separate locations for Africans under the Natives (Urban Areas) Act, initially passed in 1923 and frequently amended thereafter. Enforcement was backed by legislation such as the Native Laws Amendment Act of 1937 which gave sweeping powers to magistrates over African rights in the cities, including the right to own and occupy land. Restrictions and finally prohibitions on the African right to purchase land outside specifically designated areas were introduced culminating in the Native Land and Trust Act of 1936. Restrictions on the Indian right to purchase property were increased, resulting in the racial zoning of residential areas in a number of towns and cities under the Asiatic Land Tenure and Indian Representation Act of 1946. Within the older parts of the cities the Slums Act of 1934, although ostensibly providing for the implementation of urban redevelopment schemes, was effectively operated as a means of enforcing segregation through the demolition of integrated areas and the rehousing of those displaced in segregated municipal housing estates (Parnell 1988). Despite these measures, the resultant 'segregation city' retained pockets of integration and a large number of interspersed segregated housing estates (Davies 1981) (Figure 13.1).

In 1948 the National Party came to power, pledged to pursue a policy of strict apartheid. Thus after 1948 town planning schemes were systematically designed to establish totally segregated cities (Christopher 1990). Particular attention was devoted to those remaining areas of racial integration often dating from colonial times. They were to be removed and their inhabitants rehoused in segregated suburbs.

To this end the entire population was classified into discrete groups under the Population Registration Act passed in 1950. Effectively four groups were adopted:

> *Whites*, people of European origin, but excluding those of mixed origin;
> *Africans*, people indigenous to Africa, south of the Sahara, but excluding the Khoisan population;
> *Asians*, people of Asian origin, mainly Indians, with the exception of the descendants of slaves imported by the Dutch in the seventeenth and eighteenth centuries (Cape Malays);
> *Coloureds*, all other people. In consequence this group was highly diverse in origins and background.

Periodically other groups such as Chinese and Cape Malays were recognised, where the numbers in a particular town were deemed sufficient to establish another community. Elaborate physical and linguistic means were adopted to determine an individual's race, usually with the purpose of excluding from the White group those with some measure of mixed ancestry. The official terminology employed changed periodically as, for example, indigenous Africans were successively renamed Natives, Bantu and finally Blacks.

In the same year, 1950, the Group Areas Act provided that separate group areas were to be set aside in every town and city where only those defined as members of that group might reside. The creation of the mechanisms to achieve the aim of total segregation took a number of years to put in place and removals took place in stages. In the 1950s large numbers of Africans were resettled in the peripheral townships, while in the 1960s and 1970s a substantial proportion

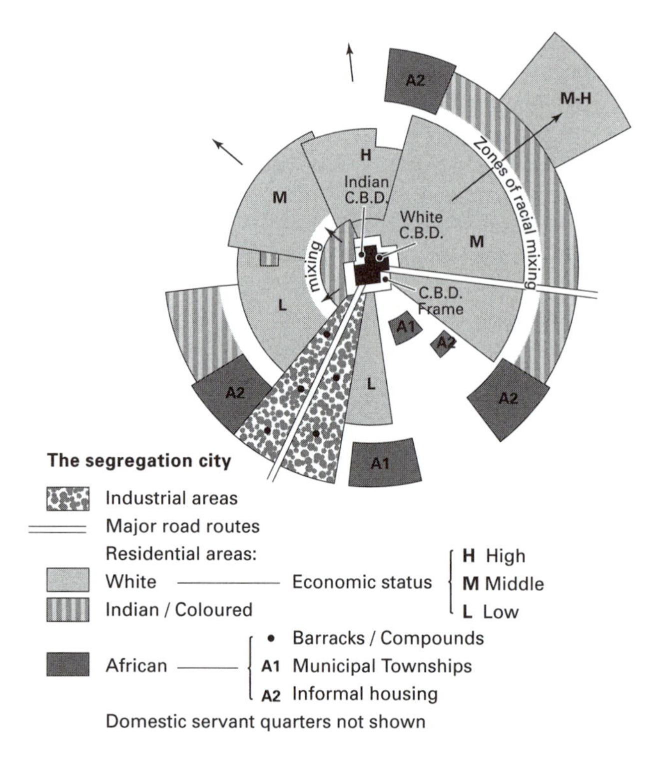

Figure 13.1 Models of the segregation and apartheid cities
Source: After Davies (1981)

of the Coloured and Indian communities was similarly evicted and resettled. Symbolically, in Johannesburg the inner, predominantly African, suburb of Sophiatown was redeveloped as the White suburb of Triomf, while in Cape Town the predominantly Coloured suburb of District Six was replanned as White Zonnebloem.

The apartheid city

The apartheid city thus developed a peculiar and distinctive structure and form (Davies 1981, Western 1981). The city centre and the inner suburbs were zoned for White occupation (Figure 13.1). Around this a series of broad sectors were drawn. The more extensive, low-density, residential areas were reserved for the White population. Thus most of the city area already built up at the time of the proclamation of the group areas was zoned for Whites (Figure 13.2). Furthermore, Coloured and Indian sectoral areas were separated from them by buffer strips of open waste land, designed to reduce social contact between members of the different groups. The Indian and Coloured group

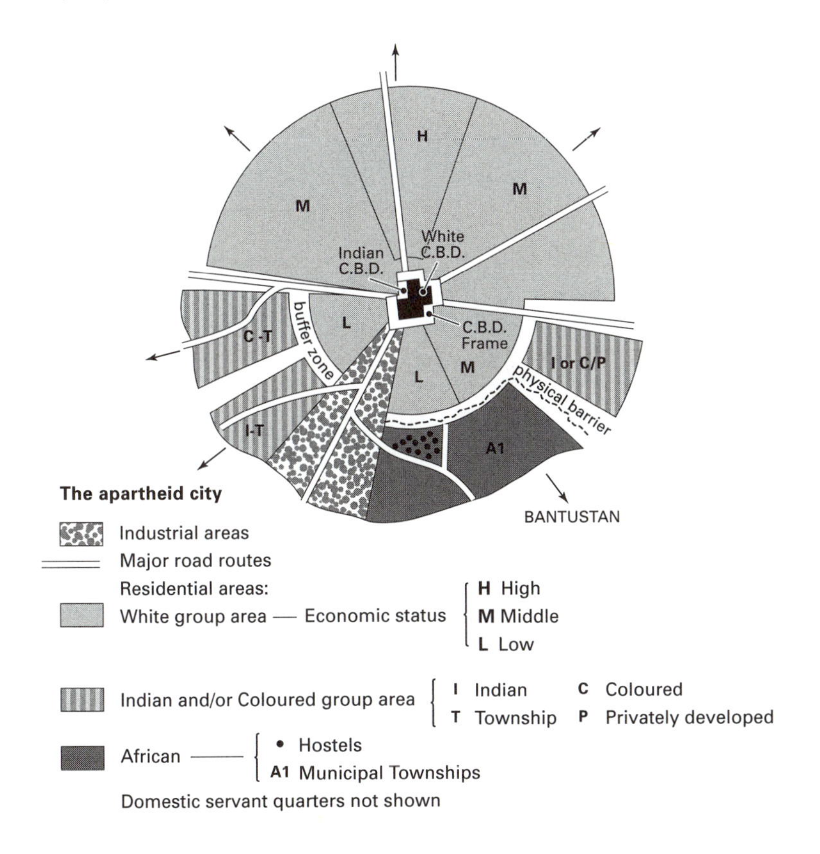

areas were generally dominated by government housing schemes dating from the era of the implementation of apartheid, with only small areas of pre-existing housing included. The broad sectoral plan of the apartheid city conflicted with the more narrowly focused plan of the segregation city. Thus a large number of segregated municipal housing estates, mainly for Coloureds and Africans, which had been built before 1948, did not conform to the broad sectoral plan of the apartheid city. Their residents were resettled and the area either demolished or the houses reoccupied by members of the entitled group.

Yet more peripheral were the African townships or locations, of variable ages, but predominantly dating from the 1950s and 1960s. Africans were unable to obtain secure freehold title as they were regarded as 'temporary sojourners' in the cities, which were viewed as essentially part of 'the domain of the White Man'. Industrial areas and lines of communication were planned as part of the overall apartheid scheme to enhance segregation and increase the social and physical distance between the various groups. By the 1970s the development of adjacent African homelands was incorporated into the apartheid city plan as long-distance commuting was introduced as a means of further increasing segregation (Krige 1988).

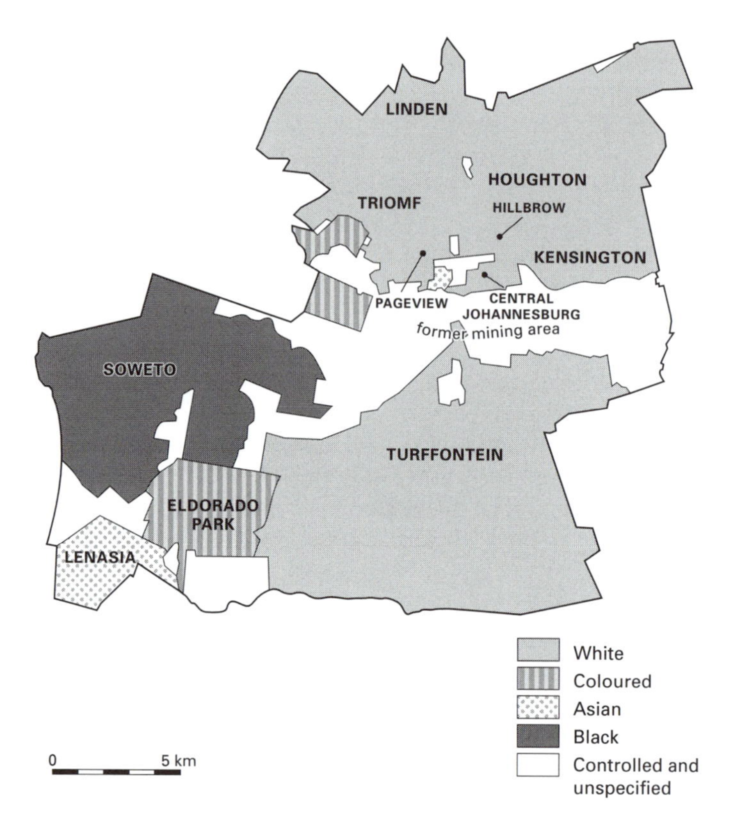

Figure 13.2 Group areas in Johannesburg, 1985

Source: Johannesburg municipality

Segregation levels thus rose dramatically during the apartheid era as integrated areas were physically destroyed and people who did not conform to the broad plan were evicted and resettled elsewhere. Most segregation indices reached over 90 by the time of the census in 1991, and were still rising (Table 13.1). Some 91.4 per cent of the enumerated urban population of the country in 1991 resided in conformity to the apartheid model. If the under-enumeration of the African population is taken into account, the proportion was probably yet higher. Those who did not conform were predominantly African servants and labourers living on their employers' properties. Furthermore, because of the racial, but not linguistic, homogeneity of most African areas, it made little difference to the calculation of segregation indices that the African areas of most cities were regarded for bureaucratic ease as a single enumeration district in the 1991 census (Christopher 1994b). Segregation between linguistic groups in the African urban areas was also pursued, but with less effect than between Whites and other groups (Christopher 1989).

Table 13.1 White indices of segregation for the major metropolitan areas of South Africa, 1951–91

Metropolitan	*White index of segregation*				
	1951	*1960*	*1970*	*1985*	*1991*
Johannesburg	75	82	89	96	96
Cape Town	71	78	88	94	96
Durban	71	78	88	89	91
Pretoria	72	80	84	93	88
Port Elizabeth	79	89	94	96	98

Source: Calculated from the records of the Central Statistical Service, Pretoria

It is noticeable that the creation of the apartheid city was planned and organised over a long period of time. Although the Group Areas Act was passed in 1950, group areas were only proclaimed on any scale after 1958, when the mechanisms to expropriate, evict and resettle people on a large scale had been put in place (Christopher 1991). Opposition to the Act was sustained, but ultimately ineffective, as the government forcibly removed people from areas deemed to be White (Platzky and Walker 1985).

The full force of the programme fell initially upon the African population with the destruction of inner-city locations and the expropriation of African freehold properties in the 1950s. The Resettlement of Natives Act of 1954 provided the immediate means to remove Africans who had been able to obtain freehold property before the prohibitions on African ownership had been introduced. At the same time, the majority of the African locations or townships near city centres were removed (Pirie and Hart 1985). In addition Africans living on White-owned properties were subject to increased restrictions and removals. Even African domestic servants were the subject of government attention. Those living in the central flatlands of Johannesburg were, for example, removed in an effort to 'whiten' the areas concerned (Mather 1987).

The laws were systematically applied and extensive new African townships were built on the periphery of the metropolitan and other urban areas. The spate of building from the mid-1950s to the late 1960s effectively transformed the structure and appearance of South African cities (Morris 1980). The standardised housing form and mass construction processes resulted in a remarkable uniformity across the country. One of the significant aspects of the rebuilding was the emergence of the extensive barrack or hostel complexes for single workers (Pirie and da Silva 1986). In official circles migrant labour was favoured over permanent workers in an attempt to pursue the idea of 'White cities' not dependent upon permanent African labour and to prevent workers' families from moving to the cities. The conditions which emerged in the hostels were usually extremely poor, leading to high levels of deprivation and ultimately conflict (Penderis and van der Merwe 1994).

The planning of the metropolitan areas was linked to that of the African homelands. In the case of cities such as Durban and Pretoria, sections of homelands were within close proximity to the city. The government's border industrial programme was designed to locate industry close to the borders of the homelands in order to make use of African labourers and workers, who would otherwise have had to migrate to the 'White' towns had the industry been located in a pre-existing industrial area. In some cases the African locations and townships were excised from the existing city and administratively transferred to the adjacent homeland. Thus the KwaMashu township in Durban was transferred to the control of the KwaZulu homeland. In other cases new suburbs were built within the homeland boundaries to house workers who would commute on a daily basis to nearby factories in the 'White' area. An important example was GaRankuwa in the Bophuthatswana homeland, near Pretoria. With Bophuthatswana's grant of 'independence' in 1977, the daily journey to work became 'international'! In some metropolitan areas the distances involved were considerable, necessitating bus journeys of several hours per day. A substantial proportion of the African population of the Bloemfontein metropolitan area was resettled at Botshabelo some 60 km from the city.

Inevitably restrictions on rural–urban migration resulted in misplaced urbanisation. Those seeking jobs in the metropolitan areas, particularly the Pretoria–Witwatersrand and Durban regions, migrated to sections of adjacent homelands and created informal settlements on land offered by the local homeland authorities. The result was the emergence of extensive squatter slums in the Wintersveld and around Durban (Soni and Maharaj 1991).

Changes in the late apartheid city

Although the dominant trend in the apartheid city after 1948 was towards greater levels of segregation, counter factors leading to integration began to take effect in the 1980s. The most significant was the pressure exerted by the shortage of accommodation in the African, Indian and Coloured group areas. Thus people moved back to live in areas where they were legally prohibited from doing so. In 1982 in a Transvaal High Court ruling in the Govender case it was specified that the government had to provide alternative accommodation for people evicted under the Group Areas Act (Morris 1994). Owing to the chronic shortage of accommodation outside the White areas this was virtually impossible to undertake.

Prior to 1982 the government's usual reaction to the institution of a legal constraint upon the implementation of its programme had been to pass another piece of legislation blocking the loophole (Carr 1990). However, delicate negotiations were in progress to co-opt leaders of the Coloured and Indian populations into a new constitutional dispensation. This culminated in the establishment of the tricameral Parliament, inaugurated in 1984, with separate legislative houses to represent the White, Coloured and Indian populations.

Political pressures exerted by the Coloured and Indian people were thus sufficient to prevent the loophole in the Group Areas Act from being blocked. Thereafter there was a limited in-migration of Indians, Coloureds and Africans to the central city and inner suburbs, which were being abandoned by Whites as they migrated to the outer suburbs. The level of in-migration was such that the ethnic composition of some previously White areas changed dramatically. Thus the suburb of Mayfair in Johannesburg became predominantly Indian (Fick, de Coning and Olivier 1988). In addition members of excluded communities could apply for permits to reside in the White areas. Initially (late 1970s) this concession to the apartheid model was applied to the Chinese community, later to the Indian and Coloured populations and finally to Africans (Nel 1993). In all these cases the numbers of families involved was relatively small, although usually concentrated in specific areas.

The extent of racial and ethnic mixing increased, particularly in the central city flatlands, and the government sought to restrict its areal extent. Initially it did this through the re-zoning of a number of central areas, for example Mayfair in Johannesburg and Woodstock in Cape Town, for Indian and Coloured occupation respectively. Significantly, this undermined the broad sectoral model of the apartheid city. However, re-zoning did not solve the problem of unrestricted integration and the government sought to confine mixing by introducing yet another category of land. In 1988 the Free Settlement Areas Act was passed which provided for the proclamation of 'grey' areas where anyone might live. Few such areas had been proclaimed by mid-1991 when the whole system was abolished.

In 1984 the government passed legislation recognising the permanency of the African population in South African cities. This recognition had far-reaching consequences as it represented a complete reversal of earlier attitudes. This was followed in 1986 by the repeal of the legislation restricting the migration of Africans to the urban areas – the infamous pass laws (Giliomee and Schlemmer 1985). As a result there was a substantial influx of poor Africans from the rural areas and poverty-stricken homelands. Most migrants could not afford formal housing and so informal housing or shack and squatter settlements expanded rapidly on the peripheries of towns and on open land within the existing African townships. Furthermore, many people who had been living, often illegally, in cramped conditions in the existing African areas sought a place of their own (Crankshaw and Hart 1990). The result was a massive expansion of designated African areas around the metropolitan regions. Particularly significant was the extension of the Cape Town metropolitan area with the designation of Khayelitsha for Africans. Previously the Cape Town metropolitan area had been declared to be part of the Coloured Preference area from which Africans were systematically excluded in order to protect Coloured labour (Cook 1986).

Surprisingly, the levels of segregation in the late apartheid city continued to rise. Public attention had often been focused upon the inner suburbs, where newspaper reporters and university academics tended to be concentrated. The continuing processes of segregation, notably for the large numbers of African migrants to the cities, were still firmly in operation. New peripheral suburbs were almost entirely segregated, so that the substantial growth of population within the cities and metropolitan areas continued to be housed within a framework established by the apartheid planners. However, the emerging patterns of the late apartheid city possibly provide the best guide to the future form of the post-apartheid city (Simon 1989).

The post-apartheid era

Since June 1991 all restrictions on land ownership and occupation by race have been removed from the statute book. Some changes have taken place, although no recent population figures are available to quantify the movements. However, a number of significant factors inhibit major changes. First, the African population is generally extremely poor. Urban unemployment rates of approximately 50 per cent indicate the scale of the economic disaster which has befallen this society (Riordan 1988). The majority of urban Africans are thus unable to afford to move into any type of formal housing. Most Africans who are in employment occupy relatively junior or menial positions and are unable to afford middle- or upper-class residences, which are mainly to be found in the former White group areas. Second, the White population has remained in South Africa and so the major urban migrations which followed decolonisation elsewhere on the continent have not been replicated. Thus there was no sudden transformation of residential patterns, although the successful political transformation has led to heightened expectations among the previously disadvantaged communities.

If there was no sudden economic transformation, there have been significant changes in urban patterns. The trends discernible in the late apartheid city accelerated. First, African numbers have continued to grow rapidly as migrants have sought work in the urban areas and urbanisation levels have risen substantially. Those living in restricted backyard shacks have sought to establish their own houses on independent plots. The result has been the reduction of densities in the African areas, as outward expansion of the informal housing areas has gathered pace. Second, inner city transformation has accelerated as increasing numbers of Africans have obtained incomes sufficient to afford central city flats and houses, close to their place of work. The emergence of high-density districts of varying qualities has been the result. Finally, suburban integration has occurred on a limited basis as those wishing to leave the African, Coloured and Indian townships, and able to afford the costs, have sought properties in the previous White area.

Change is apparent, but there is a remarkable lack of statistical information to quantify the movements taking place. Even the censuses are officially believed to have at least a 20 per cent undercount (South Africa 1992a). In practice it is probable the actual figures in some cities could be substantially more. On such an imperfect statistical base it is foolhardy to be too dogmatic. However, it is proposed to examine the situation in Port Elizabeth as a case study of the implementation and impact of segregation in South African metropolitan areas.

Port Elizabeth

The current (1995) population of Port Elizabeth is estimated to be over 1 million, of whom approximately 150,000 are Whites; 160,000 Coloureds; 8,000 Indians and 700,000 Africans. This represents a very different racial composition from that recorded in 1951 at the beginning of the apartheid era, when the population only totalled 200,000, of whom 40 per cent were White and 35 per cent were Africans.

In 1951 the city was already markedly segregated as a result of the various segregation practices dating from colonial times. It is notable that the first separate urban location for the indigenous population in South Africa was established in Port Elizabeth by the London Missionary Society as early as 1834. The peculiar characteristics of the population, with a mixture of Western Cape features, including a substantial Coloured population, overlapped those of the eastern parts of the country with its African majorities. In addition, the population of the Cape Colony Province enjoyed a colour-blind political franchise and the relative freedom of all the people to purchase and occupy property until the 1930s. This resulted in the appearance of residential segregation based on economic status rather than purely on a racial basis, as is evident in the three eastern provinces. However, the establishment of formal segregated African locations was pursued in parallel with residential integration, although only half the African population lived in the New Brighton municipal township in 1951 (Figure 13.3).

The imposition of a group areas plan upon the city in 1961 led to a substantial restructuring of its physical layout and population distribution. The broad sectoral plan established by the Group Areas Board provided that the southern and western sectors were zoned White and the northern sectors were zoned for Indians, Coloureds and Africans. A separate group area for the Chinese community was also laid out. As the city was to be reordered some 31 per cent of the total population was subject to resettlement (Table 13.2). Virtually none of the White population was moved, but 48 per cent of the remainder was subject to removal. Thus substantial evictions and resettlements took place, resulting in the destruction of extensive housing and business areas where everything was demolished and subject to comprehensive redevelopment. The most significant such areas were South End, Salisbury Park and Fairview within the southern White sector. As in other cities, the movement of Africans preceded the imposition of the group areas plan. Thus African removals from freehold properties in the inner suburbs and from Korsten resulted in a substantial expansion of New Brighton, which tripled its population in the course of the

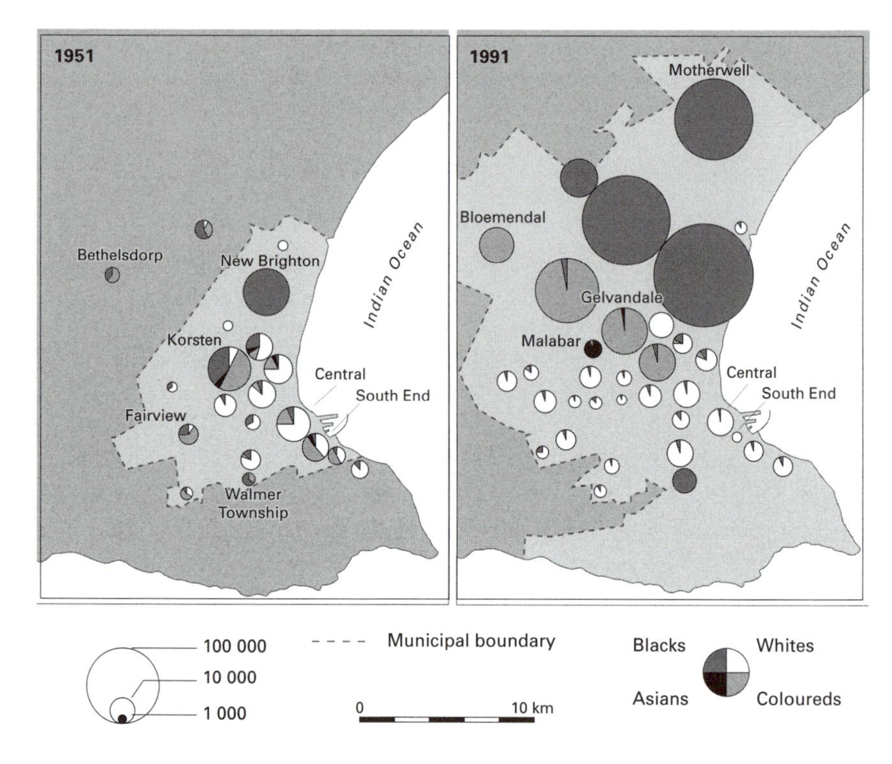

Figure 13.3 Distribution of population in Port Elizabeth, 1951–91

Source: Christopher (1994a)

1950s, leaving only 12 per cent of the African population outside the townships by 1960.

In the northern sectors, new segregated suburbs were built with extensive tracts of waste land between them to act as buffer strips. It is noticeable that the White area, which was the most important area of official concern, recorded only a modest increase in population between 1951 and 1991, despite substantial suburban developments, whereas the population in the other, more peripheral group areas increased dramatically. By 1991 the residence of some 97 per cent of the city's population conformed to the apartheid plan (Table 13.3).

Changes since the repeal of the Group Areas Act have not been as extensive as the release of built-up pressures might lead one to expect. The outward growth of African informal housing areas has been most noticeable. Schemes to reduce the densities in the most overcrowded areas have resulted in African housing being developed in areas previously zoned for other groups. As yet there have been no land invasions, such as have taken place in Durban and Johannesburg. In part this reflects the availability of large tracts of municipal and church lands for low-cost or informal housing development and the highly

Table 13.2 Population distribution in Port Elizabeth, 1951

Area	*Numbers resident in areas as eventually zoned*				
	Whites	*Coloureds*	*Asians*	*Africans*	*Total*
White	79,719	23,579	3,317	15,016	121,631
Coloured	725	17,274	724	15,466	34,189
Asian	0	0	0	0	0
African	166	3,560	88	40,450	44,264
Total	80,610	44,413	4,129	70,932	200,084

Source: Compiled by the author from the records of the Central Statistical Service, Pretoria

visible attempts by the transitional authorities to address the needs of the poor (Goduka and Riordan 1994). Group areas buffer strips, public open spaces and nature reserves offer additional vacant land close to the city centre and comparatively inexpensive connection to municipal services. Major site-and-service housing schemes have been commenced with the aim of reducing the markedly high population densities (de-densification) in the former African areas (Table 13.4). Significantly, the initial schemes have been located in the former Coloured group areas, another major break in the apartheid plan. Within the former African areas major upgrading schemes have commenced, aimed at providing services to shack areas.

Within the city centre, flats, even offices converted to bedrooms, have been rented to all race groups, although in some blocks all the new lessees are from groups previously excluded. Occupation policy is the preserve of the owner as the constitutional provisions on combating racial discrimination in the private sector are as yet uncertain. A predominantly African and Coloured residential area has developed adjacent to the central business district, paralleling the experience of central Johannesburg. Elsewhere integration has been piecemeal, with few marked patterns evident. Within the previously White areas, new housing schemes appear to be more integrated than the older segregated areas. This applied to both the upper economic end of the market, notably Summerstrand,

Table 13.3 Population distribution in Port Elizabeth, 1991

Group area	*Resident in group areas*				
	Whites	*Coloureds*	*Asians*	*Africans*	*Total*
White	126,628	3,252	898	5,860	136,638
Coloured	206	125,565	1,322	1,362	128,455
Asian	11	409	5,163	185	5,768
African	16	4,567	2	339,986	344,571
Total	126,861	133,793	7,385	347,393	615,432

Source: Compiled by the author from the records of the Central Statistical Service, Pretoria

Table 13.4 Extent of group areas and population, Port Elizabeth, 1991

Group	*Population*	*Group area (ha)*	*Persons per ha*
White	137,381	13,757	10.0
Coloured	153,646	10,062	15.3
Asian	8,449	672	12.5
African[a]	345,972	4,811	71.9

Source: South Africa (1992b) and records of the Department of Provincial Affairs, Pretoria

Note
a African population was subject to substantial underenumeration.

and the lower end at Fairview, marking a return to the pattern of segregation based on economic and social status predating apartheid (South Africa 1992b).

The disparities in income levels between the population groups are striking. Whereas only 27 per cent of employed Whites in Port Elizabeth earned less than R 10,000 (£2,000) per annum in 1991, some 78 per cent of Africans did so. On the basis of the total urban population some 27 per cent of Whites earned over R 30,000 (£6,000) per annum, compared with 0.8 per cent of Africans. If the estimated 60 per cent African unemployment rate is also considered, the economic restraint upon any significant African migration to the previous White areas will be appreciated.

The question of land claims may partially change the situation. The newly constituted Commission on Restitution of Land Rights and the Land Claims Court are charged with the task of restoring property expropriated by the state under racially based laws, such as the Group Areas Act. Some areas such as Salisbury Park and the northern two-thirds of Fairview have lain waste since the time of the evictions and so can be claimed individually by the previous owners or their heirs. Other areas such as South End, where an all-White suburb and government buildings were erected obliterating the property lines of the old plan, cannot be reclaimed and so compensation either in cash or alternative land must suffice. Large cash payments are beyond the limited resources of the new government. Thus the situation of the alternative land will be of crucial significance in determining the reintegration process. It would appear that the Port Elizabeth Land and Community Restoration Association, which has been formed to guard the interests of those dispossessed, will opt for a community-based scheme attendant upon the redevelopment of Salisbury Park and Fairview.

Port Elizabeth thus faces all the significant urban problems of the post-apartheid era. Population growth partially dependent upon the continuing migration of people from the country to the towns shows little likelihood of declining for some time. The process of de-densification will be essential to achieve acceptable standards of housing and service provisions. The provision of serviced land is therefore going to be the key to solving the mass housing crisis and restoring lost property rights. Where that land is located and how rapidly it can be delivered will determine the city's future.

Conclusion

South African cities have been subject to rigid legislated segregation on a racial basis. Cities were designed and built to reinforce segregation between the races, notably to separate the White population from the remainder of the citizens. Such areas of integration as had developed in the nineteenth and early twentieth centuries were systematically destroyed in the apartheid era in the quest to effect total segregation. However, the goal of total separation eluded the apartheid planners even after forty years of sustained effort. National government priorities changed as the need to attain greater security necessitated compromises on the subject. However, the late exceptions to the apartheid model should not blind the reader to the highly effective implementation of the policy and the sweeping transformation of South African cities and metropolitan areas which were effected in its pursuit.

The repeal of the legislative basis of the apartheid city does not mean that integration will follow, as economic class was closely allied to racial group. At present the majority of the African population is too poor to enter the formal housing market. However, as economic advancement takes place, so greater racial integration will become possible. It seems probable that the inherited physical structures of the apartheid city are likely to preclude rapid integration. Further, it appears that without a significant rise in living standards among the previously excluded communities, the active restructuring of the physical characteristics of South African cities will be slow. The national Reconstruction and Development Programme thus has a vital role to play in changing South African society and with it the form of the cities and metropolitan areas.

14

WELFARE STATE EFFECTS ON INEQUALITY AND SEGREGATION

Concluding remarks

Herman van der Wusten and Sako Musterd

Cities by their nature provoke a wide variety of sentiments. Urban life has been ecstatically praised and deeply loathed. In the literary world and in those places where current opinion is crafted, such sentiments continuously struggle for pride of place. In the social sciences efforts have been made over the years to assess the state of the urban question. Some have emphasised the glories of urbanity, others have concentrated on the miseries of life in the city, and some have propounded a balanced view through an appreciation of both sides of the coin simultaneously.

In the current debate about cities much attention is given to global cities as dominating centres of the planet-wide interlinked economy, from which cultural innovations arise and are diffused. At the same time these bustling places are portrayed as concentrations of vice, deprivation and loneliness. And these two views are integrated: on account of their global role, which provides them with these valuable qualities, cities are also of necessity confronted with the flip side of the coin. Obviously this invites the question of whether politics (local, national, any level) can retain and support the advantages and mitigate the disadvantages.

The authors of this book have mainly been concerned with the worrying aspects of urban existence. Some of them have taken into account the forces of globalisation that bring at least some (but by no means all) cities into a strong position within the world-wide economic system. In the very special case of South Africa one can also argue that globalisation has undermined the capacity of the apartheid regime to seal itself off from the outside world in a sustainable fashion. But whatever the advantages, cities have in any case to cope with the difficulties that international interdependencies engender. Employment structures have profoundly changed. Seemingly stable welfare state regimes (and also those within cities) that for some time have softened social problems have become strained. Culturally different migrants have arrived in their numbers, particularly in cities, and new problems of mutual accommodation have arisen.

The cases in this book deal with the current state of the urban question in western welfare states with a special view to spatial segregation and exclusion. Welfare states are by no means all of a kind. A major interest in this project has in fact been what difference a specific type of welfare state makes when it has to deal with the new conditions under which the urban question crops up. Under the new condition of intensifying globalisation, welfare states in general seemingly have more difficulty than they had to keep state societies intact, that is, to help avoid significant sections of society no longer being able to make meaningful social contact with the rest of society. This is because welfare states have fewer resources at their disposal for these purposes and new population categories have arrived that are more prone to become such segregated groups. There is, of course, the counter-argument that welfare states have themselves induced this falling apart of state societies by lavish spending on certain population categories that are therefore tempted to withdraw from social life in general. By and large we reject such claims. In so far as welfare state arrangements have indeed enabled people to withdraw from society, the prime causes of these processes are usually elsewhere. At the same time welfare state regimes have, of course, to be financed; consequently, they are only sustainable at levels that the state of the economy prescribes and employment allows, and employment is one of the main instruments that enables people to cross divides within society. Thus, welfare state regimes have a double interest in promoting employment levels.

Cities are apparently the places where these problems, however caused and now often labelled as 'exclusion', come most insistently to the fore. Exclusion refers to the prevention or frustration of social relationships. Welfare state regimes have mainly operated on the assumption that the unjust distribution of assets (income, health, education, etc.) is the prime mover of such broken relationships and that consequently distribution should be more just. Just is not the same as equal, but huge inequalities go against the grain of opinions about justified differences in all welfare state systems. Spatial segregation within cities might be an important condition of exclusion: although such a link has been more often proposed than demonstrated, demonstrations pertain to too restricted a set of cases to allow generalisation and there are counter-arguments. In addition, spatial segregation comes in different guises (small scale versus large scale, with respect to different kinds of segmentation like income, ethnicity, lifestyle) and these differences are crucial when it comes to consequences.

In dealing with segregation and exclusion within cities under the condition of intensifying globalisation, and thus reductions in the autonomy of welfare state regimes, we have to be aware of two additional factors. First of all, welfare states formed major barriers to segregation and exclusion when they were, for one brief generation after the Second World War, the hegemonial project of state formation in the western world. But the different nature of welfare state arrangements across the western world affected the precise nature of those barriers to segregation and exclusion. By no means always intentionally, the position of public housing in the housing stock, the way it was organised, the allocation rules, and more indirectly the redistribution of incomes, had an impact on the socio-spatial differentiation of the urban residential population and consequently on segregation patterns. Compare, for example, the different outcomes of segregation in Sweden and Belgium in this volume, in part based on the different nature of the housing system. In the same way, the whole array of welfare provisions diminished unfair differences and thus the likelihood of social exclusion. Think, for example, of the different ways in which health insurance and education have been organised across the western world. Second, on a wider historical canvas, cities cannot fruitfully be considered as random chunks of territory where, through a momentary conjunction of factors, some social formation develops. They have a historical position within a state and within wider urban networks that is part of the conditions of the present. This position importantly affects their ability to perform in a globalising world and thus colours the roles of major protagonists. See in this volume the traditional special role of Paris as the capital city of the French state and the implications for policy priorities.

For some time the differentiation of urban residential populations and their dynamics could usefully be portrayed along three dimensions: economic or social status, ethnic status, stage in the family cycle (Shevsky and Bell 1955). On the assumption of fairly straightforward patterns of household formation, households could be classified along these three dimensions, according to which they tended to be concentrated in different parts of cities, creating an urban mosaic (Timms 1971). As their position changed they moved, or if the composition of a neighbourhood population changed collectively, the position of the neighbourhood in the urban housing market was altered. Now, household structures in western societies have become far less unilinear and homogeneous: different patterns of employment within households plus welfare state systems have generally unravelled the direct link between occupation and income at the household level, although this was always liable to be destabilised by the existence of independent wealth in certain pockets of society. Ethnic status has lost its 'natural' character, too, on account of quickly changing population composition.

We should stress that social area analysis maps in fact always concentrated on the spatial differentiation of social categories whose substantive mutual differences were taken for granted. In fact, the importance of gaps in social status dependent on occupation and ethnicity, and of the differences in lifestyle that accompany stages in the life cycle, will vary across societies and over time. Accordingly, following our earlier argument, the chances of exclusion based on different general distributions of assets will differ. Weak substantive differences will on the face of it be less likely to lead to sharp segregation, but a word of caution is in order. Objectively weak substantive differences may be considered large enough for people directly involved to prefer residential segregation. Consequently, it is not impossible for objectively weak substantive differences and pronounced segregation to go together. Sharp substantive differences may more often result in sharp segregation but this is not necessarily the case, at least not on all spatial levels. A good example of mix despite sharp discrepancies along all dimensions is the occupation of some of the Haussmannian apartment buildings along major boulevards in Paris, where for a long time singles, often of foreign origin and with slender means, resided in the old service quarters on the upper floors and in the attics while bourgeois families lived in the huge flats from the first floor up. Thus, the importance of differences within the residential population does not necessarily coincide with levels of spatial segregation along the same dimension at commonly used levels of aggregation.

In this collection household structure has only a walk-on part and our authors have concentrated on occupational, more generally socio-economic, status and ethnic status. For the sake of simplicity (and disregarding the extra sources of status differentiation deriving from education, occupation, independent wealth, etc.) we further refer to income inequality to indicate substantive, unjust differences and to segregation according to income to indicate varying residential distributions between income categories. In the same way we will refer to ethnic inequality and ethnic segregation. When either income or ethnic status differences are very pronounced, exclusion is supposed to follow. In the case of segregation opinions are more divided. It depends on the spatial scales involved: the larger the units the higher the probability of exclusion. A major reason why this seems convincing is the provision of a self-sufficient environment within larger units with no incentives to use urban space at large. This, of course, is not only connected with spatial scale. It also has to do with what is available in the immediate environment and the extent to which people are allowed out (remember militia-controlled urban neighbourhoods in civil war situations like Beirut).

All the authors in this book deal with the socio-economic dimension of segregation and exclusion in cities, though some of them only tangentially. There are definitely differences in this respect between these cities in terms of inequality as well as segregation. Chicago and Port Elizabeth may be the extremes at one end of the scale while Hamburg, Stockholm and Amsterdam are at the opposite end. The welfare state regime apparently makes a difference. The USA and South Africa qualify only marginally as welfare states at best, though for different reasons. On the other hand, Sweden has for a long time been the clearest expression of the social-democratic welfare state regime in action, while Hamburg is a city long governed by social democrats, in a federal state with a prodigious welfare state system that has been called conservative for its differentiated, but extensive entitlements, and Amsterdam is embedded in a hybrid but again extensive welfare state system with one of the largest public housing sectors in the western world (that in Amsterdam itself has seen its finest hour). The cases in between represent cities in welfare state systems of intermediate quality.

Nevertheless, various aspects of the impact of welfare state regimes on socio-economic inequality and segregation are less than obvious or are even obscure. Two examples: the two British case studies each show the relevance of the welfare state regime but they do not point in the same direction. Whereas the case of London (chapter 2) demonstrates the impact of globalisation on income inequality, the British welfare state regime, even under the former Conservative government, performs a mitigating role in comparison with the impact of globalisation in the USA. The case of Edinburgh (chapter 7) shows the degradation of the British welfare state regime in the field of public housing during the same period and the concomitant increase in socio-economic segregation. The cases of Amsterdam and Stockholm (chapters 11 and 12) tell the stories of two cities embedded in two of the most elaborate welfare state regimes ever, particularly in the field of housing. The concomitant low levels of socio-economic inequality and segregation, however, do not give rise to an absence of social exclusion and modest levels of political concern. Even the relatively minute degree of socio-economic inequality and the gentle level of segregation in Amsterdam give rise to empirical indications of different rates of social exclusion dependent on address, whereas very slight increases in levels of socio-economic inequality and segregation in Stockholm are a problem of serious policy concern in Sweden. In conclusion, welfare state regimes are multi-dimensional and they do not necessarily change in all respects in the same direction at the same time. They also each create their own reality and scale of evaluation. International comparison may elucidate the differences between those scales.

The great majority of our cases also deal with ethnic inequality and segregation. Generalising the distinction introduced in chapter 3 between long-established ethnic differences and those between a settled population and new immigrants, we have in fact three cases where these long-established differences loom large: Chicago (chapter 4), Port Elizabeth (chapter 13) and Belfast (chapter 6). They can usefully be compared to the cases of Toronto (chapter 5), Brussels (chapter 8), Paris (chapter 9) and Hamburg (chapter 10), where the emphasis is on recent immigrants. However serious the incorporation of the new immigrants may be, in terms of the intractability of the social problems emanating from these long-standing ethnic distinctions, they are still much less serious.

The cases of Chicago, Port Elizabeth and Belfast are of course significantly different. They include the heritage of an enslaved and legally discriminated-against population that arrived in an urban market-driven society, the heritage of an enslaved or indentured and then unprecedentedly discriminated-against population in the urban environment itself, and the heritage of two opposing nationalisms deriving their respective strengths from a deep religious divide.

These cases demonstrate convincingly that globalisation is not the only, not even a necessary, driving force for the production of extremely serious social problems related to cultural distinction in current cities. The ethnic differences are the products of earlier processes, perhaps earlier rounds of globalisation, but distinctly different from the current round. Welfare state regimes have not had the chance to demonstrate their mitigating power in these instances. To the extent that they have operated, they have had to deal with fairly intractable situations. The short-lived effort to achieve the Great Society floundered in Chicago, as in many other places in the USA; the apartheid regime lacked the minimum credentials of a liberal democratic regime that are part and parcel of the current definition of a welfare state (it would still be important to look at apartheid also as an authoritarian form of a minimum welfare state); the British take-over of the Northern Irish government after the troubles started in 1969 has belatedly introduced large-scale welfare provision in the province, but has not resulted in more than an uncertain cessation of violence between the contending forces of nationalism despite the fact that a large part of the population is more than tired of the conflict.

What these cases have in common is ingrained cultural difference perceived in a hierarchical manner. As a consequence there are recurring reminders from voices on both sides of the contested nature of a shared society from which exclusion, and also withdrawal, has become frequent. Welfare provisions cannot flourish on this basis. The cultural divide in these cases tends to be further deepened by socio-economic inequality. A hierarchical division of labour ensues. The cultural divide that also tends to result in cultural segregation will in addition become a pattern of advanced socio-economic segregation. From the perspective of the minimisation of exclusion this is a situation to be avoided at all costs.

The solution is an all-encompassing concept of citizenship that informs policy over the longer period and/or the granting of a certain autonomy within spheres of public policy like housing, health care and education within an overall framework of agreement and financed from the public purse on a proportional basis. The combination of these two policy lines has, sometimes with difficulties, preserved civil peace in the consociational democracies of Europe and enabled the construction of extensive appropriate welfare state regimes.

The case of Belfast, and Northern Ireland more generally, is instructive in this respect. The expansion of housing provision under welfare state regime rules has in this case coincided with an increase in culturally based segregation. Diminishing ethnic inequality, at least in the eyes of the governing powers, perhaps even generally diminishing income inequality, has been accompanied by sharpening ethnic segregation. This seems to go in the direction prescribed by those in favour of consociational democracy institutions in this type of situation. In the political realm several extensive efforts in that direction have been made, so far to no avail. The charged history of the conflict has prevented the mobilisation of sufficient trust to overcome the cultural divide. This does not mean that it is the wrong road, but merely indicates how hard it is.

The distinction just made between old and new immigrants to the city is, of course, quite relative. In many of the other cities a major part of the population also stems from immigration at some stage in the past. Most of these immigrants, however, no longer show up emphatically as such after a few generations. This implies that cities and their governments have a window of opportunity – in policy terms a pretty large window of perhaps a few generations – in which new immigrant populations have to be accommodated in such a way that cultural divides do not harden in a hierarchical manner and that the chances of social exclusion are minimised. Welfare state regimes are the political constructions evolved over the years as a result of many contradictory processes that are meant to prevent social exclusion. If neglected by mainstream politics in the different countries and cities of Europe, the problems of Chicago, Port Elizabeth and Belfast may well spread. New political claims elsewhere may also be built on the basis of a cultural divide by, on the one hand, a combination of 'autochthonous', culturally inspired nationalists and defenders of the welfare status quo against perceived new invaders (the new extreme/radical right) and, on the other, a congeries of political interests of migrant groups from straightforward interest articulation via dreams of local secessions to diaspora politics that link them to political forces in the home country and that very often have a senselessly radical flavour. The Hamburg case (chapter 10) shows the early signs of such a process. The long-term effort by Parisian and French national authorities to move lower-class people, many immigrants among them, from central Paris into the *grands ensembles* at the periphery of the Parisian agglomeration has contributed to the emergence of radical immigrant politics and the Front National.

Some degree of spatial segregation of new, lower-class immigrants is inevitable and not simply to be assessed as negative. New immigrants may benefit from each other and a spatial concentration of immigrants may better sustain supporting institutions. From the perspective of the authorities ameliorative policy can more easily be targeted if there is spatial concentration. Some spatially concentrated communities may act as springboards to overcome the forces that push in the direction of exclusion. Others may become refuges for a population of those who have failed or withdrawn from society at large. A long time ago the distinction between slums of hope and slums of despair was introduced and again and again similar notions crop up in the literature. The clearest case in this collection is the analysis of two neighbourhoods in Brussels, both populated by immigrants and apparently on a different trajectory in this respect. A major policy and research problem all over the countries of the European Union is the extent to which policy directed to the prevention of exclusion should be spatially targeted, what the target should be (what size, what to include and exclude in terms of policy) and how the degradation of neighbourhoods can be stopped and the positive developments in other deprived

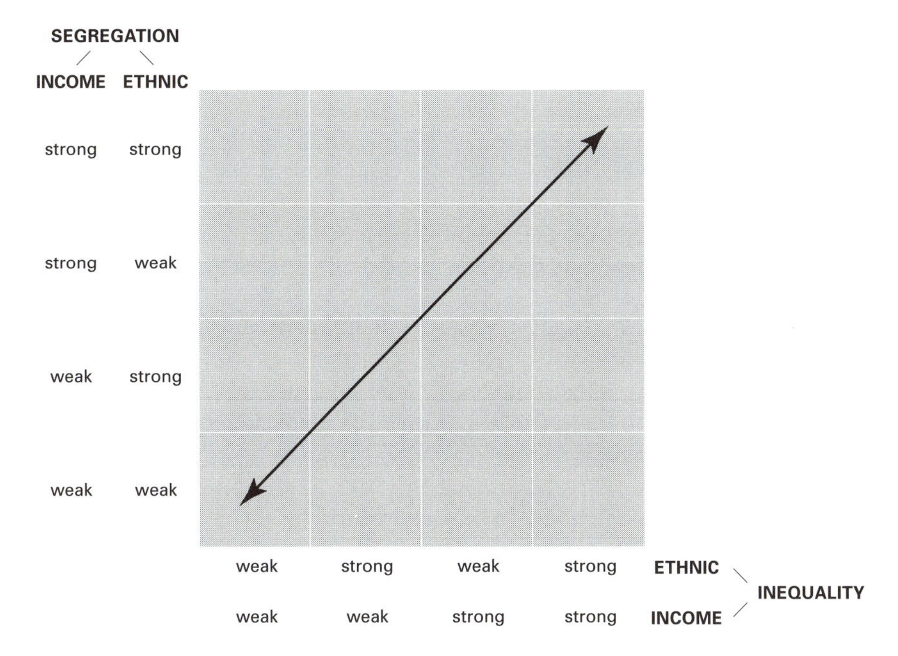

Figure 14.1 Typology of cities according to segregation and inequality

neighbourhoods can be supported. In a series of urban projects all over the Union experiments are now under way. They should, in addition, be carefully assessed with an eye to the administrative structures and welfare state policy regimes in which they have to function.

This book, though held together by a common perspective, cannot pretend to be a truly comparative venture. There are formidable barriers to such a project, if only from the perspective of data acquisition. An earlier effort to come up with comparable segregation indices for a number of the European cities analysed in this volume showed that differences in spatial units and in the definition of ethnic categories make it hard, though not entirely impossible, to produce comparative data (Musterd, Ostendorf and Breebaart 1997). Much more work with respect to income and its corollaries also has to be done.

In order to see at least an outline of what such a comparative effort might yield, we construct the typology that has been implicitly used in the preceding remarks and speculate on some of its implications. By making dichotomous distinctions of both types of inequality and segregation and enumeration of all the ensuing combinations, we get a sixteen-fold typology along two dimensions. Figure 14.1 shows the result.

In general, the expectation is that welfare state regimes push cities towards the lower left corner. To the extent that their welfare programmes contain more universal entitlements of more important size there will be a push to the left. If they encompass an important public housing component, the terms for a softening effect on segregation levels improve: all other things being equal, there will be a downward pull. Welfare state regimes as a rule do not formally differentiate between categories with different ethnic status, though there is some historical precedent in the way the arrangements in some of the consociational democracies came about and there is perhaps a grudging admission of ethnic status in the way Britain currently incorporates the nationalist-informed Catholic/Protestant divide in Northern Ireland in its welfare provisions in the province.

Despite these few cases to the contrary, overall this suggests that the distinctions in the typology in terms of ethnic segregation and ethnic inequality are useless for the purpose of tracing welfare state influence in urban segregation and exclusion. However, there may be ethnically differentiating consequences of specific welfare policies (e.g. due to ethnically differential access and informal discrimination) and there may be compensatory gestures towards deprived ethnic categories (e.g. protected quotas). The consequences for the overall distribution may be modest, but the social consequences may nevertheless be important (compare the discussions on the emergence of a middle class of Black Americans and the consequences for the remaining Blacks in the ghettos (Wilson 1987)).

Globalisation and its apparent impact on welfare state regimes, employment structures and incoming new immigrants would tend to push cities to the upper right-hand corner. Income inequalities increase through changes in the composition of the job supply and/or through changes in the number of available jobs. The composition of the levels at which the new immigrants enter the labour market determines a major part of the ethnic inequality. The more they are concentrated at the lower end, the more ethnic inequality increases. The more job opportunities at the lower end of the labour market collapse, jobs are offered at lower remuneration and social security does not compensate for the losses, while there is an unimpeded rise of salaries at the higher end of the scale, the more unequal the income distribution becomes. As public housing is confined to fewer segments and fewer areas, or public housing can no longer compete with housing offered in other segments of the housing market for a diverse public, and new immigrants congregate by discrimination or preference (according to circumstances) or both, income and ethnic segregation will increase.

As in our view inequality and segregation do not necessarily go together, it is intriguing to speculate where cities are positioned off the diagonal or do not move along this path. In some chapters the emphasis is on one of the two dimensions (e.g. segregation in Paris, income inequality in London), so that we cannot know. It looks as if Belfast moves in an unusual direction: towards less income inequality but at the same time more pronounced levels of segregation, that is to the upper left corner in our figure. That is the consequence of the imposition of a welfare state regime akin to those earlier introduced in the consociational democracies that helped overcome culturally divided societies elsewhere in Europe. Other cases with the most pronounced levels of segregation do indeed coincide with high levels of inequality: the USA and South Africa. There are thus different ways of promoting segregation: through following the preferences of a deeply divided population by administrative regulation (Northern Ireland), through the translation of the same kind of preferences by the 'free' market (USA), by imposition of a regime itself (South Africa). If South Africa is able to move in the direction of a more conventional welfare state regime, will ethnic and/or income segregation be reduced before ethnic and/or income inequality have weakened (top to bottom) or the other way round (right to left) or inevitably together or in different directions depending on whether we consider the ethnic or the income dimension?

How should exclusion be minimised? In contemporary metropolitan areas there seems to be a powerful trend for like to assort with like. People prefer to be among their own kind, or the structure of the housing market offers homogeneous blocks of residences with specific residential environments adopted to this homogeneous residential population, or both. Even if absolute differences are not so strong, this implies a sharply edged urban mosaic at the upper left corner of our figure. There are, of course, parts of the urban structure where more mixed populations live, if not together at least in the same neighbourhood. Within gentrified neighbourhoods economic inequality may of course be pronounced. We move towards the lower right-hand corner of the figure. Gentrification notwithstanding, this is generally not the normal urban situation nowadays. In several chapters in this book the fear is that western societies are all moving to the right-hand upper corner and this results in higher rates of exclusion. The Amsterdam chapter suggests that, even at very modest levels, inequality and segregation both give rise to diminishing participation and thus exclusion. If this is regarded as a situation to be avoided, policy makers should do what they can to keep their cities in the lower left quadrant of the figure or to bring them there.

One final remark concerning the units of analysis in this collection should be made. While the main perspective here has been about the impact of globalising forces on the internal functioning of cities as mediated by welfare state regimes, it is interesting that hardly any attention has been paid to local government. The demise of the nation-state may be on many academic research agendas, possibly to be superseded by transnational networks of cities; the view in this collection has generally been that the nation-state, through its institutions, rules and policies, still sets the terms in which many of the aspects of city life can best be studied. It is much too early to do away with the nation-state altogether despite the intensive restructurings that it has undergone over the last twenty years or so. Welfare state regimes may have lost their aura of inevitable expansion for ever, but they survive in recognisable form.

REFERENCES

African National Congress (1994) *The Reconstruction and Development Programme*, Johannesburg: Umayano.

Amersfoort, H. van (1992) 'Ethnic residential patterns in a welfare state: lessons from Amsterdam, 1970–1990', *New Community* 18, 3: 439–56.

Andersen, E. (1990) *Streetwise: Race, Class, and Change in an Urban Community*, Chicago: University of Chicago Press.

Andersson, R. (1996) 'The geographical and social mobility of immigrants: escalator regions in Sweden from an ethnic perspective', *Geografiska Annaler Serie B* 78B, 1: 3–25.

Andersson, R. and Molina, I. (1996) *Etnisk Boendesegregation i Teori och Praktik – En Internationell Överblick och en Nationell Lägesbeskrivning*, Bil-SOU:55, Stockholm: Fritzes.

Arend, Michael (1984) 'Segregation zwischen Schweizern und Ausländern in der Stadt Zurich', *Dokumente und Informationen zur Schweizerischen Orts-, Regional- und Landesplanung* 20: 31–5 (DISP No. 75).

Atkinson, A.B. (1993) 'What is happening to the distribution of income in the UK', *LSE Welfare State Programme*, WSP 87.

Axelsson, S. and Fournier, A. (1993) 'The shift from manufacturing to services in Sweden', *Urban Studies* 30: 285–98.

Badcock, B. (1984) *Unfairly Structured Cities*, London: Blackwell.

Balakrishnan, T.R. and Hou, F. (1995) *The Changing Patterns of Spatial Concentration and Residential Segregation of Ethnic Groups in Canada's Major Metropolitan Areas 1981–1991*, London, Ontario: Population Studies Centre, University of Western Ontario, Discussion Paper no. 95–2.

Banting, K. (1987) 'The welfare state and inequality in the 1980s', *Canadian Review of Sociology and Anthropology* 24: 309–38.

Bennett, L. (1993) 'Harold Washington and the black urban regime', *Urban Affairs Quarterly* 28, March: 423–40.

Bergmann, Eckhard and Peters, Aribert (1994) 'Auslander und Wirtschaft', *Informationen zur Raumentwicklung* 5/6: 387–98.

Bernow, R. and Strömqvist, U. (1995) *Nya Bostäder i Stockholm? Vilkoren för Bostadsbyggande i Stockholms Stad 1995–2000. Utredning*, Stockholm: TEMAPLAN AB.

Biterman, D. (1994a) 'Boendesegregationsutveckling i Stockholms län 1970–1990', mimeo, Stockholm: Inregia AB.

— (1994b) 'Utveckling i miljonprogramområdena i Stockholms län 1970–1990', mimeo, Stockholm: Inregia AB.

Boal, F.W. (1987) 'Segregation', in M. Pacione (ed.) *Social Geography: Process and Prospect*, London: Croom Helm, pp. 90–128.

— (1995) *Shaping a City: Belfast in the Late Twentieth Century*, Belfast: Institute of Irish Studies, The Queen's University of Belfast.

Boal, F.W., Doherty, P. and Pringle, D.G. (1974) *The Spatial Distribution of Some Social Problems in the Belfast Urban Area*, Belfast: Northern Ireland Community Relations Commission.

Boal, F.W., Keane, M.C. and Livingstone, D.L. (1996) *Them and Us?: A Survey of Catholic and Protestant Churchgoers in Belfast*, Belfast: Report to Northern Ireland Office Central Community Relations Unit.

Body-Gendrot, Sophie (1994) 'Violently divided cities: a new theme in political science', in Seamus Dunn (ed.) *Managing Divided Cities*, Keele, Staffordshire: Ryburn Publishing/Keele University Press, pp. 214–27.

Borgegård, L.-E. and Murdie, R. (1994) 'Social polarization and the crisis of welfare state: the case of Stockholm', *Built Environment* 20: 257–68.

Bourdieu, Pierre (1984) *Die feinen Unterschiede*, 3rd edn, Frankfurt/M.: Suhrkamp.

Bourne, L.S. (1992) 'Population turnaround in the Canadian inner city', *Canadian Journal of Urban Research* 1: 66–89.

— (1993a) 'The demise of gentrification? A commentary and prospective view', *Urban Geography* 14: 95–107.

— (1993b) 'Close together and worlds apart: an analysis of the changes in the ecology of income in Canadian cities', *Urban Studies* 30: 1293–1317.

— (1994) 'The role of gentrification in the changing ecology of income: evidence from Canadian cities and implications for future research', in G. O. Braun (ed.) *Managing and Marketing of Urban Development and Urban Life*, Berlin: Dietrich Beimer Verlag, pp. 561–74.

Bovenkerk, F., Gras, M.J.I. and Ramsoedh, D. (1995) *Discrimination against Migrant Workers and Ethnic Minorities in Access to Employment in the Netherlands*, Geneva: International Labour Office.

Boverket (1993) *Svensk Bostadsmarknad i Internationell Belysning. Årsbok 1993 från Boverket*, Stockholm: Fritzes.

Brechet, R. (1994) 'Plans et schémas directeurs en Ile-de-France', *Cahiers de l'IAVRIF* 108: 49–61.

Brown, P. and Crompton, R. (eds) (1994) *Economic Restructuring and Social Exclusion*, London: UCL Press.

Browning, R.P., Marshall, D.R. and Tabb, D.H. (1990) 'Has political incorporation been achieved? Is it enough?', in R.P. Browning, D.R. Marshall and D.H. Tabb (eds) *Racial Politics in American Cities*, New York: Longman, p. 243.

Buck, N., (1994) 'Social divisions and labour market change in London: national, urban and global factors', unpublished paper for ESRC London Seminar, 28 October 1994.

Buck, N., Drennan, M. and Newton, K. (1992) 'Dynamics of the metropolitan economy', in S.S. Fainstein, I. Gordon and M. Harloe (eds) *Divided Cities: New York and London in the Contemporary World*, Oxford: Blackwell, ch. 3.

Built Environment (1994) 20, 3, special issue (A rising European underclass?).

Canada Employment and Immigration Centre (1992) *Economic Outlook: Toronto Region*, Toronto: Canada Employment and Immigration Centre, Toronto Regional Economist Office.

Carpenter, J. (1994) 'Urban policy and social change in two Parisian neighbourhoods, 1962–1992', unpublished Ph.D. thesis, University of Sheffield.

Carr, W.J.P. (1990) *Soweto: Its Creation, Life and Decline*, Johannesburg: South African Institute of Race Relations.

Castel, R. (1995) *Les Métamorphoses de la question sociale*, Paris: Fayard.

Castles, S. (1993) 'Population distribution and migration', in *Proceedings of the United Nations Expert Meeting on Population Distribution and Migration*, Santa Cruz, Bolivia, pp. 309–33.

Castles, S. and Miller, M.J. (1993) *The Age of Migration. International Population Movements in the Modern World*, London: Macmillan Press Ltd.

Caulfield, J. (1994) *City Form and Everyday Life: Toronto's Gentrification and Critical Social Practice*, Toronto: University of Toronto Press.

Charef, M. (1983) *Le Thé au harem d'Archi Ahmed*, Paris: Mercure de France. (The film, with Charef as the director, was released in 1985.)

Chicago Department of Housing (1994) *Comprehensive Housing Affordability Strategy*, Chicago: Department of Housing.

Chicago Department of Planning and Development (1993) *Population by Race and Latino Origin for Census Tracts, Community Areas, and Wards 1980, 1990*, Chicago: Department of Planning and Development.

Chicago Rehab Network (1993) *The Chicago Affordable Housing Fact Book: Visions for Change*, Chicago: Chicago Rehab Network.

Chicago Tribune (1985) *The American Millstone: An Examination of the Nation's Permanent Underclass*, Chicago and New York: Contemporary Books.

Christopher, A.J. (1988) 'Roots of urban segregation: South Africa at Union 1910', *Journal of Historical Geography* 14: 151–69.

—— (1989) 'Apartheid within apartheid: an assessment of official intra-Black segregation on the Witwatersrand, South Africa', *Professional Geographer* 41: 328–36.

—— (1990) 'Apartheid and urban segregation levels in South Africa', *Urban Studies* 27: 421–40.

—— (1991) 'Changing patterns of group area proclamations in South Africa, 1950–1989', *Political Geography Quarterly* 10: 240–53.

—— (1992) 'Urban segregation levels in the British Overseas Empire and its successors in the twentieth century', *Transactions of the Institute of British Geographers* (New Series) 17: 95–107.

—— (1994a) *The Atlas of Apartheid*, London: Routledge.

—— (1994b) 'Segregation levels in the late-apartheid city 1985–1991', *Tijdschrift voor Economische en Sociale Geografie* 85: 15–24.

Citizenship and Immigration Canada (1994a) *Facts and Figures: Overview of Immigration*, Hull, Quebec: Strategic Research, Analysis and Information Branch, Policy Sector, Citizenship and Immigration Canada.

—— (1994b) *A Broader Vision: Immigration and Citizenship, Plan 1995–2000, Annual Report to Parliament*, Hull, Quebec: Citizenship and Immigration Canada.

City of Edinburgh (1979) Edinburgh District Council, Housing Committee, *Annual Report 1980–1981.*

City of Toronto Planning and Development Department (1984) *Toronto Region Incomes.* Research Bulletin 24.

Coffey, W. (1994) *The Evolution of Canada's Metropolitan Economies*, Montreal: The Institute for Research on Public Policy.

Coleman, James S. (1986) 'Social theory, social research, and a theory of action', *American Journal of Sociology* 91: 1309–35.

Commission of the European Communities (1993) *European Social Policy: Options for the Union: Green Paper*, Luxembourg: Office for Official Publications of the European Communities.

—— (1994) 'Proposal for a Council Decision adopting a specific programme in the field of targeted socio-economic research (1994–1998)', Ref: 94/0091 (CNS).

—— (1995) *Targeted Socio-Economic Research Programme (1994–1998) – Workprogramme*, Brussels.

Committee of Planning and Coordinating Organizations (1992) *A Social Report for Metro*, Toronto: Social Planning Council of Metropolitan Toronto.

Community Training and Research Services (1993) *Poverty amongst Plenty: Surveys of*

Taughmonagh and Clarawood Estates, Belfast: Community Training and Research Services.
Cook, G.P. (1986) 'Khayelitsha – policy change or crisis response?', *Transactions of the Institute of British Geographers* (New Series) 11: 57–66.
Coyne, A. (1994) 'Keeping up: the high price of looking after our own', *The Globe and Mail*, 15 October.
Crankshaw, O. and Hart, T. (1990) 'The roots of homelessness: causes of squatting in the Vlakfontein settlement south of Johannesburg', *South African Geographical Journal* 72: 65–70.
Cross, Malcolm and Waldinger, Roger (1992) 'Migrants, minorities, and the ethnic division of labor', in Susan S. Fainstein, Michael Harloe and Ian Gordon (eds) *Divided Cities*, Oxford: Blackwell, pp. 151–74.
Daalen, G. van, Deurloo, R. and Musterd, S. (1995) *Segregatie op Microniveau*, Amsterdam: AME, Universiteit van Amsterdam.
Danermark, B. (1983) *Klass, Inkomst och Boende: om Segregation i Några Kommuner*, Örebro: Högskolan i Örebro.
Dangschat, Jens S. (1994) 'Concentration of poverty in the landscapes of "boomtown" Hamburg: the creation of a new urban underclass', *Urban Studies* 31: 1133–47.
—— (1995) '"Stadt" als Ort und als Ursache von Armut und sozialer Ausgrenzung', *Politik und Zeitgeschichte* B 31–2: 50–62.
Davies, James C. (1962) 'Toward a theory of revolution', *American Sociological Review* 27: 5–19.
Davies, R.J. (1981) 'The spatial formation of the South African city', *GeoJournal* Supplementary Issue 2: 59–72.
Davies, W.K.D. and Murdie, R. (1993) 'Measuring the social ecology of cities', in L.S. Bourne and D. Ley (eds) *The Changing Social Geography of Canadian Cities*, Montreal and Kingston: McGill-Queen's University Press, ch. 3.
Davis, Mike (1990) *City of Quartz*, London: Verso.
De Decker, P. (1994) 'Onzichtbare muren, over leven in achtergestelde buurtne en de reproductie van sociale uitsluiting', *Planologisch Nieuws* 14, 4: 341–66.
De Keersmaecker, M.L. (1997) 'Indicateurs de la situation sociale à Bruxelles', paper presented at the 'Vingtième anniversaire des CPAS – Colloque de la Commission Communautaire Commune', Brussels, 12 March 1997.
De Lannoy, W. and Kesteloot, C. (1990) 'Het scheppen van sociaal-ruimtelijke ongelijkheden in de stad', in Werkgroep Mort-Subite, *Barsten in België*, Berchem: EPO, pp. 143–78.
Denton, N.A. and Massey, D.S. (1988) 'Residential segregation of Blacks, Hispanics and Asians by socio-economic status and generation', *Social Science Quarterly* 69: 797–818.
Department of Finance, Canada (1995) *Budget Plan, February 27*, Ottawa: Department of Finance.
Deurloo, M.C. (1987) *A Multivariate Analysis of Residential Mobility*, Utrecht: Knag.
Dieleman, F. and Hamnett, C. (1994) 'Globalisation, regulation and the urban system', *Urban Studies* 31: 357–64.
DN (1993) *Inkomstklyftan. En Artikelserie om Stor-Stockholm i Dagens Nyheter 1993*, Stockholm: DN.
Doherty, P and Poole, M.A. (1995) *Ethnic Residential Segregation in Belfast*, Coleraine: Centre for the Study of Conflict, University of Ulster.
Drake, S. and Cayton, H. (1945) *Black Metropolis: A Study of Negro Life in a Northern City*, New York: Harcourt Brace.
Dreier, P. and Hulchanski, J.D. (1993) 'The role of nonprofit housing in Canada and the United States: some comparisons', *Housing Policy Debate* 4: 43–81.
Economist (30 July 1994) 'Europe and the underclass', pp. 19–21.

Eisinger, P. (1980) *The Politics of Displacement*, New York: Academic Press.

Employment and Immigration Canada (1992) *Immigration Statistics, 1991*, Ottawa: Supply and Services Canada.

Engbersen, G. (1990) *Publieke Bijstandsgeheimen. Het Ontstaan van een Onderklasse in Nederland*, Leiden/Antwerp: Stenfert Kroese.

Erbe, Brigitte M. (1975) 'Race and socioeconomic segregation', *American Sociological Review* 40: 801–12.

Esping-Andersen, G. (1989) 'The three political economies of the welfare state', *The Canadian Review of Sociology and Anthropology* 26: 10–36.

— (1990) *The Three Worlds of Welfare Capitalism*, Cambridge: Polity Press.

— (1993) *Changing Classes: Stratification and Mobility in Post-Industrial Societies*, London: Sage.

Fainstein, Norman S. (1993) 'Race, class, and segregation: discourses about African Americans', *International Journal of Urban and Regional Research* 17: 384–403.

— (forthcoming) 'Black ghettoization and social mobility', in Michael P. Smith and Joe Feagin (eds) *The Bubbling Cauldron*, Minneapolis: University of Minnesota Press.

Fainstein, Susan S. and Fainstein, Norman (1989) 'The racial dimension in urban political economy', *Urban Affairs Quarterly* 25, December: 187–99.

— (forthcoming) 'Urban regimes and black citizens: the economic and social impacts of black political incorporation in US cities', *International Journal of Urban and Regional Research.*

Fainstein, S., Gordon, I. and Harloe, M. (eds) (1992) *Divided Cities: New York and London in the Contemporary World*, Oxford: Blackwell.

Fainstein, Susan S. and Harloe, Michael (1992) 'Introduction: London and New York in the contemporary world', in Susan S. Fainstein, Ian Gordon and Michael Harloe (eds) *Divided Cities*, Oxford: Blackwell, pp. 1–28.

Fegelman, A. (1991) 'Suburbs more integrated, but have a long way to go', *The Chicago Tribune*, 2 February 1991.

Fick, J., de Coning, C. and Olivier, N. (1988) 'Ethnicity and residential patterning in a divided society: a case study of Mayfair in Johannesburg', *South Africa International* 19: 1–27.

Financial Post (1993) *The Financial Post 500: Canada's Largest Corporations*, Toronto: Financial Post.

Fong, E. (1994) 'Residential proximity among racial groups in U.S. and Canadian neighborhoods', *Urban Affairs Quarterly* 30: 285–97.

Forrest, R. and Murie, A. (1990) *Moving the Housing Market*, Avebury.

— (1995) 'From privatisation to commodification: tenure conversion and new zones of transition in the city', *IJURR* 19, 3: 407–22.

Forrest, R., Murie, A. and Gordon, D. (1995) *The Resale of Former Council Homes*, London: Her Majesty's Stationery Office.

Freie und Hansestadt Hamburg, Behörde für Arbeit, Gesundheit und Soziales (1993) *Armut in Hamburg*, Hamburg: Behörde für Arbeit, Gesundheit und Soziales.

— (1994) *Hamburger Arbeitsmarktbericht (1991–1993/94)*, Hamburg: Behörde für Arbeit, Gesundheit und Soziales.

Friedmann, J. and Wolff, G. (1982) 'World city formation: an agenda for research and action', *International Journal for Urban and Regional Research* 4: 309–43.

Friedrichs, Jürgen (1985) 'Ökonomischer Strukturwandel und Disparitäten von Qualifikationen', in Jürgen Friedrichs (ed.) *Die Städte in den 80er Jahren*, Opladen: Westdeutscher Verlag, pp. 48–69.

Friedrichs, Jürgen, Jagodzinski, Wolfgang and Dülmer, Hermann (1994) 'Städtische Konflikte und Wahlverhalten', *Hamburg in Zahlen* 1: 36–41.

Fritzell, J. and Lundberg, O. (1994) *Vardagens Vvillkor; Levnadsförhållanden i Sverige under Tre Decennier*, Stockholm: Brombergs.

Fulton, J. (1995) 'Ethnicity and state form in the division of Ireland', *New Community* 21, 3: 341–55.
Funken, K. and Cooper, P. (eds) (1995) *Old and New Poverty. The Challenge for Reform*, London: Rivers Oram Press.
Gad, G. (1985) 'Office location dynamics in Toronto: suburbanization and central district specialization', *Urban Geography* 6: 331–51.
Gans, H. (1990) 'Deconstructing the underclass: the term's dangers as a planning concept', *American Planning Association Journal* 56, 3: 271–7.
Gardiner, K. (1993) 'A survey of income inequality over the last twenty years – how does the UK compare?', *London School of Economics Welfare State Programme*, WSP 100.
Gemeente Amsterdam (1993) 'Sociaal isolement zeer ten dele verklaard door achterstand', *Op grond van Cijfers* 1: 1–3.
Giles, W. and Preston, V. (1991) 'Ethnicity, gender and labour markets in Canada: a case study of immigrant women in Toronto', paper presented at the annual meetings of the Canadian Association of Anthropology and Sociology.
— (1996) 'The domestication of women's work: a comparison of Chinese and Portuguese immigrant women homeworkers', *Studies in Political Economy* 51: 147–81.
Giliomee, H. and Schlemmer, L. (1985) *Up against the Fences: Poverty, Passes and Privilege in South Africa*, Cape Town: David Philip.
Ginsburg, N. (1992) *Division of Welfare. A Critical Introduction to Comparative Social Policy*, London: Sage Publications.
Girard, A. (1977) 'Opinion publique, immigration et immigrés', *Ethnologie Française* 7: 219–28.
Glascow, D. (1980) *The Black Underclass: Poverty, Unemployment and Entrapment of Ghetto Youth*, San Francisco: Jossey-Bass.
Goduka, P. and Riordan, R. (1994) *The Rebuilding of Port Elizabeth*, Port Elizabeth: African National Congress.
Goering, J. (1992) 'Towards the comparative exploration of public housing segregation in England and the United States', *Housing Studies* 8: 256–73.
Goldberg, M. and Mercer, J. (1986) *The Myth of the North American City*, Vancouver: University of British Columbia Press.
Gordon, I. and Sassen, S. (1992) 'Restructuring the urban labour markets', in S.S. Fainstein, I. Gordon and M. Harloe (eds) *Divided Cities: New York and London in the Contemporary World*, Oxford: Blackwell, ch. 4.
Gramsci, A. (1971) *Selections from the Prison Notebooks*, London: Lawrence & Wishart.
Green, H. and Booth, P. (1996) 'Six years of urban policy in France: the "contrat de ville" programme', *Modern and Contemporary France* 4: 183–92.
Grillo, R. (1985) *Ideologies and Institutions in Urban France: The Representation of Immigrants*, Cambridge: Cambridge University Press.
Grimes, Seamus (1993) 'Residential segregation in Australian cities: a literature review', *International Migration Review* 27: 103–20.
Grofman, Bernard N. and Muller, Edward N. (1973) 'The strange case of relative gratification and potential for political violence: the V-curve hypothesis', *American Political Science Review* 67: 514–39.
Guglielmo, R. and Moulin, B. (1986) 'Les grands ensembles et la politique', *Hérodote* 43: 39–74.
Gurr, Ted R. (1970) *Why Men Rebel*, Princeton, NJ: Princeton University Press.
Hague, C. (1993) 'The restructuring of the image of Edinburgh: the politics of urban planning in a European Arena', *Proceedings of the 9th Urban Change and Conflict Conference*, University of Sheffield.
Halleröd, B. (1994) 'Poverty in Sweden: a new approach to the direct measurement of consensual poverty', *Umeå Studies in Sociology*, Umeå: Umeå University.

— (1996) 'När har Sverige blivit nog ojämnlikt? De svenska hushållens ekonomi 1985 och 1992', in *Ekonomisk Debatt 1996* 24, 4: 267–79.

Hamnett, C. (1979) 'Area based explanations: a critical appraisal', in D.T. Herbert and D.M. Smith (eds) *Social Problems and the City: Geographical Perspectives*, Oxford: Oxford University Press, pp. 244–60.

— (1984) 'Housing the two nations: socio-tenurial polarisation in England and Wales 1961–81', *Urban Studies* 43–9: 387.

— (1994a) 'Social polarisation in global cities: theory and evidence', *Urban Studies* 31: 401–24.

— (1994b) 'Socio-economic change in London: professionalisation not polarisation', *Built Environment* 20: 192–203.

— (1996a) 'A stroke of the Chancellor's pen: the social and regional impact of the Conservatives' abolition of higher rates of income taxation', *Environment and Planning A*, 1407–30.

— (1996b) 'Social polarisation, economic restructuring and welfare state regimes', *Urban Studies* 8: 1407–30.

Harloe, M. (1995) *The People's Home? Social Rented Housing in Europe and America*, Oxford: Blackwell.

Harloe, M., Marcuse, P. and Smith, N. (1992) 'Housing for people, housing for profits', in S.S. Fainstein, I. Gordon and M. Harloe (eds) *Divided Cities*, Oxford: Blackwell, pp. 175–202.

Harrison, B. and Bluestone, B. (1988) *The Great U-Turn: Corporate Restructuring and the Polarizing of America*, New York: Basic Books.

Hart, D.M. (1988) 'District Six – political manipulation of urban space', *Urban Geography* 9: 603–28.

Harvey, D. (1973) *Social Justice and the City*, Oxford: Basil Blackwell.

Heinstedt, L. (1992) *Boendemönster som Kunskapskälla. En Studie av Allmännyttan i Stockholms Län*. Arbetsrapport. Gävle: Statens institut för byggnadsforskning.

Heisler, Barbara Schmitter (1994) 'Housing policy and the underclass: the United Kingdom, Germany, and the Netherlands', *Journal of Urban Affairs* 16, 3: 203–20.

Henry, F. (1994) *The Caribbean Diaspora in Toronto: Learning to Live with Racism*, Toronto: University of Toronto Press.

Hermalin, Albert I. and Farley, Reynolds (1973) 'The potential for residential integration in cities and suburbs: implications for the busing controversy', *American Sociological Review* 38: 595–610.

Herrnstein, Richard J. and Murray, Charles (1994) *The Bell Curve*, New York: Free Press.

Hiebert, D. (1994) 'Canadian immigration: policy, politics, geography', *The Canadian Geographer* 38: 254–8.

Hine, J.P. and Wang, Y.P. (1994) *Change in the City: Planning and Social Change in the City of Edinburgh 1981–91*, paper given to the Societies in Transition Conference, Edinburgh, June.

Hjarnø, J. (1996) *Global Cities in Two Ways. A Comment on Saskia Sassen's Global City Hypothesis*, Esbjerg: South Jutland University Press.

Howell, C. (1992) 'The dilemmas of post-Fordism: socialists, flexibility, and labour market deregulation in France', *Politics and Society* 20, 1: 71–100.

Hulchanski, J.D. (1993) *Barriers to Equal Access in the Housing Market: The Role of Discrimination on the Basis of Race and Gender*, Toronto: Research Paper 187, Centre for Urban and Community Studies, University of Toronto.

Hwang, SeanShong, Murdock, Steven H., Parpia, Banoo and Hamm, Rita R. (1985) 'The effects of race and socioeconomic status on residential segregation in Texas, 1970–80', *Social Forces* 63: 732–47.

Inregia (1997) 'Utvecklingen i Stockholms läns miljonprogramområden efter 1990', mimeo, Stockholm.

International Labour Office (1994) *World Labour Report, 1994*, Geneva: International Labour Office.
Jackson, P. (1989) *Maps of Meaning: An Introduction to Cultural Geography*, London: Unwin Hyman.
Jacquier, C. (1994) *Voyage à travers dix quartiers en crise*, Paris: L'Harmattan.
Jaynes, Gerald David and Williams, Robin M. Jr (eds) (1989) *A Common Destiny*, Washington, DC: National Academy Press.
Jencks, C. (1992) *Rethinking Social Policy: Race, Poverty, and the Underclass*, Cambridge, MA: Harvard University Press.
—— (1994) 'The homeless', *The New York Review* 21 April: 20–7.
Jencks, C. and Peterson, P. (eds) (1991) *The Urban Underclass*, Washington, DC: The Brookings Institution.
Jenkins, S.P. and Cowell, F.A (1994) 'Dwarfs and giants in the 1980s: trends in the UK income distribution', *Fiscal Studies* 15: 99–118.
Johansson, M. (1970) *Om Levnadsnivåundersökningen. Utkast till kap 1 och 2 i Betänkande om Svenska Folkets Levnadsförhållanden*, Stockholm: Allmänna Förlaget.
Judd, Dennis (1994) 'Urban violence and enclave politics: crime as text, race as subtext', in Seamus Dunn (ed.) *Managing Divided Cities*, Keele, Staffordshire: Ryburn Publishing/Keele University Press, pp. 160–75.
Kalbach, W. (1990) 'Ethnic residential segregation and its significance in an urban setting', in R. Breton, W.W. Isajiw, W.E. Kalbach and J.G. Reitz (eds) *Ethnic Identity and Equality: Varieties of Experience in a Canadian City*, Toronto: University of Toronto Press, ch. 3.
Kasarda, J. (1989) 'Urban industrial transition and the underclass', *Annals of the American Academy of Political and Social Science* 501, 1.
Kasarda, John D. and Friedrichs, Jürgen (1985) 'Comparative demographic–employment mismatches in U.S. and West German cities', in R.L. and I.H. Simpson (eds) *Research in the Sociology of Work*, vol. 3, Greenwich: JAI Press, pp. 1–30.
Kassovitz, M. (1995) *La Haine* (film).
Keane, M.C. (1990) 'Segregation processes in public sector housing', in P. Doherty (ed.) *Geographical Perspectives on the Belfast Region*, Newtownabbey: Geographical Society of Ireland, pp. 88–108.
Kelso, William A. (1994) *Poverty and the Underclass*, New York–London: New York University Press.
Kempen, E.T. van (1994) 'The dual city and the poor: social polarisation, social segregation and life chances', *Urban Studies* 31: 995–1015.
Kesteloot, C. (1994) 'Three levels of socio-spatial polarisation in Brussels', *Built Environment* 20, 3: 204–17.
—— (1995a) 'The creation of socio-spatial marginalisation in Brussels: a tale of flexibility, geographical competition and guestworkers' neighbourhoods', in C. Hadjimichalis and D. Sadler (eds) *Europe at the Margins: New Mosaics of Inequality*, Chichester: John Wiley, pp. 69–85.
—— (1995b) 'La problématique d'intégration des jeunes urbains: une analyse géographique du cas bruxellois', in C. Fijnaut (ed.) *Changement de société, crime et justice pénale en Europe*, volume I, Antwerp: Kluwer Rechtswetenschappen, I.113–I.129.
Kesteloot, C. and Van der Haegen, H. (1997) 'Foreigners in Brussels 1981–1991: spatial continuity and social change', *Tijdschrift voor Sociale en Economische Geografie*, 2: 105–19.
Kesteloot, C., De Decker, P. and Manço, A. (1997) 'Turks and housing in Belgium, with special reference to Brussels, Ghent and Visé', in S. Özüekren and R. Van Kempen (eds) *Turks in European Cities: Housing and Urban Segregation*, Utrecht: European Research Centre on Migration and Ethnic Relations, Utrecht University, pp. 67–97.

Kesteloot, C., Meert, H., Mistiaen, P., Savenberg, S. and Van der Haegen, H. (1997) *De Geografische Dimensie van de Dualisering in de Maatschappij, Overlevingsstrategieën in Twee Brusselse Wijken*, Brussels: Federale Diensten voor Wetenschappelijke, Technische en Culturele Aangelegenheden.

Kesteloot, C., Mistiaen, P. and Decroly, J.M. (1997) 'La dimension spatiale de la pauvreté à Bruxelles: indicateurs, causes et stratégies locales de lutte contre la pauvreté', in J. Vranken and K. Galloo (eds) *20 jaar OCMW in Brussel* (forthcoming).

Kesteloot, C., with Vandenbroecke, H., Van der Haegen, H., Vanneste, D. and Van Hecke, E. (1996) *Atlas van Achtergestelde Buurten in Vlaanderen en Brussel*, Brussels: Ministerie van de Vlaamse Gemeenschap.

Kilmurray, A. (1995) 'Beyond the stereotypes', in Democratic Dialogue *Social Exclusion, Social Inclusion*, Belfast: Democratic Dialogue, pp. 32–8.

Kloosterman, R. (1994) 'Three worlds of welfare capitalism? The welfare state and the postindustrial trajectory in the Netherlands after 1980s', *West European Politics* 17, 4: 166–89.

—— (1996) 'Double Dutch, polarization trends in Amsterdam and Rotterdam after 1980', *Regional Studies* 30, 5: 467–76.

Krause, Peter (1994) 'Incomes in Germany since unification: inequality, evaluation, and poverty (1990–1993)', paper presented at the XIIIth World Congress of Sociology, RC 28, Bielefeld, Germany, 18–23 July 1994.

Krige, D.S. (1988) *Die transformasie van die Suid-Afrikaanse stad*, Bloemfontein: University of the Orange Free State, Department of Town and Regional Planning.

Kuttner, B (1983) 'The declining middle', *Atlantic Monthly* July: 60–72.

Lagrée, J.-C. and Lew-Fai, P. (1989) *Jeunes et chômeurs, chômages et recomposition sociale en France, en Italie et en Grande-Bretagne*, Paris: Presses du CNRS.

Lampard, Eric E. (1986) 'The New York metropolis in transformation: history and prospect. A study in historical particularity', in Hans-Jürgen Ewers, John B. Goddard and Horst Matzerath (eds) *The Future of the Metropolis*, Berlin–New York: De Gruyter, pp. 27–100.

Lapointe Consulting Inc. and R. Murdie (1995) *Immigrant Housing Choices, 1991*, Ottawa: Canada Mortgage and Housing Corporation.

Lash, S. and Urry, J. (1993) *Economies of Signs and Spaces*, London: Sage.

Lawrence, R.Z (1984) 'Sectoral shifts and the size of the middle class', *The Brookings Review* Fall: 3–11.

Laws, G. (1994) 'Community activism around the built form of Toronto's welfare state', *Canadian Journal of Urban Research* 3: 1–28.

Lehmann, N. (1991) *The Promised Land: The Great Migration and How It Changed America*, New York: Alfred A. Knopf.

Lelévrier, C. and Pichon-Varin, F. (1995) 'Démolir les grands ensembles . . . et après?', *Habitat* 13: 17–20.

Lemon, A. (1992) *Homes Apart: South Africa's Segregated Cities*, London: Paul Chapman.

Levy, P. (1987) 'The middle class: is it really vanishing?', *The Brookings Review* Summer: 77–122.

Ley, D. (1992) 'Gentrification in recession: social change in six Canadian cities, 1981–86', *Urban Geography* 13: 230–56.

Lieberson, Stanley and Carter, Donna K. (1982) 'Temporal changes and urban differences in residential segregation: a reconsideration', *American Journal of Sociology* 88: 296–310.

Lightman, E. and Irving, A. (1991) 'Restructuring Canada's welfare state', *Journal of Social Policy* 20: 65–86.

Lipietz, A. (1989) *Choisir l'audace, une politique économique pour le XXIe siècle*, Paris: La Découverte.

LO (1996) *Rättvisa, Rättviseutredningens Rapport till LO's 23:e Ordinarie Kongress 7–12 September 1996.* Stockholm: LO.

Logan, John, Alba, Richard D. and McNulty, Thomas L. (1994) 'The racially divided city: housing and labor markets in Los Angeles', in Seamus Dunn (ed.) *Managing Divided Cities*, Keele, Staffordshire: Ryburn Publishing/Keele University Press, pp. 118–40.

Logan, John, Taylor-Gooby, Peter and Reuter, Monika (1992) 'Poverty and income inequality', in Susan S. Fainstein, Ian Gordon and Michael Harloe (eds) *Divided Cities*, Oxford: Blackwell, pp. 129–50.

Loinger, G. and Ledoux-Rehoudj, M. (1986) 'Logement et chômage: contribution à l'analyse des parcs de logement sociaux: le cas de Stains', *Villes en Parallèle* 11: 204–25.

McGregor, A. and McConnachie, M. (1995) 'Social exclusion, urban regeneration and economic reintegration', *Urban Studies* 10: 1587–1600.

McQuaig, L. (1993) *The Wealthy Banker's Wife: The Assault on Equality in Canada*, Toronto: Penguin Books.

Maguire, W.J. (1993) *Belfast*, Keele: Ryburn Publishing.

Making Belfast Work (1995) *Making Belfast Work: Strategy Statement*, Belfast: Making Belfast Work.

Mannoni, O. (1956) *Prospero and Caliban: The Psychology of Colonization*, New York: Praeger.

Marcuse, P. (1989) 'Dual city: a muddy metaphor for a quartered city', *International Journal of Urban and Regional Research* 13: 697–708.

—— (1993) 'Degentrification and advanced homelessness: new patterns, old processes', *Netherlands Journal of Housing and the Built Environment* 8, 2: 177–91.

—— (1994) 'Walls as metaphor and reality', in Seamus Dunn (ed.) *Managing Divided Cities*, Keele, Staffordshire: Ryburn Publishing/Keele University Press, pp. 41–52.

—— (1996) 'Is Australia different? Globalization and the new urban poverty', Occasional Paper 3, Melbourne: Australian Housing and Urban Research Institute.

Marshall, Harvey and Jiobu, Robert (1975) 'Residential segregation in the United States cities: a causal analysis', *Social Forces* 53: 449–60.

Marshall, J.U. (1994) 'Population growth in Canadian metropolises, 1901–1986', in F. Frisken (ed.) *The Changing Canadian Metropolis: A Public Policy Perspective*, vol. 1, Berkeley and Toronto: Institute of Governmental Studies Press, University of California and Canadian Urban Institute, ch. 1.

Massey, Douglas S. (1985) 'Ethnic residential segregation: a theoretical synthesis and empirical review', *Sociology and Social Research* 69: 315–50.

Massey, Douglas S. and Denton, Nancy A. (1985) 'Spatial assimilation as a socioeconomic outcome', *American Sociological Review* 50: 94–106.

—— (1993) *American Apartheid: Segregation and the Making of the Underclass*, Cambridge, MA: Harvard University Press.

Massey, Douglas S., Condran, Gretchen A. and Denton, Nancy A. (1987) 'The effect of residential segregation on black social and economic well-being', *Social Forces* 66: 29–56.

Mather, C. (1987) 'Residential segregation and Johannesburg's "locations in the sky"', *South African Geographical Journal* 69: 119–28.

Mayer, H. and Wade, R. (1969) *Chicago: Growth of a Metropolis*, Chicago: University of Chicago Press.

Meert, H., Mistiaen, P. and Kesteloot, C. (1997) 'The geography of survival in different urban settings', *Tijdschrift voor Sociale en Economische Geografie*, 2: 169–81.

Meinert, Jürgen (1993) 'Sozialhilfe in Hamburg im Ländervergleich 1980 bis 1990', *Hamburg in Zahlen* 1: 4–20.

Merton, Robert K. and Rossi, Alice S. (1968) 'Contributions to the theory of reference group behavior', in Robert K. Merton *Social Theory and Social Structure*, New York: Free Press, pp. 279–334.

Metropolitan Toronto Planning Department (1992) *Metropolitan Toronto Key Facts 1992*, Toronto: Metropolitan Toronto Planning Department.
— (1993a) *Metro's Changing Housing Scene, 1986–1991*, Toronto: Metropolitan Toronto Planning Department.
— (1993b) *Metro Toronto Employment Survey, Preliminary Results*, Toronto: Metropolitan Toronto Planning Department.
— (1995) *Metropolitan Toronto: Key Facts, 1995*, Toronto: Metropolitan Toronto Planning Department.
Mik, Ger (1983) 'Residential segregation in Rotterdam: background and policy', *Tijdschrift voor Economische en Sociale Geografie* 74: 74–86.
Mingione, E. (1991) *Fragmented Societies: A Sociology of Work beyond the Market Paradigm*, Oxford: Basil Blackwell.
— (ed.) (1996) *Urban Poverty and the Underclass: A Reader*, Oxford: Basil Blackwell.
Ministry of Labour (1988) *The Labour Market and Labour Market Policy in Sweden. A Discussion Paper for the 1990s*, Stockholm: Swedish Government Printing.
Mishra, R. (1990) *The Welfare State in Capitalist Society: Policies of Retrenchment and Maintenance in Europe, North America and Australia*, Toronto: University of Toronto Press.
Mistiaen, P. (1994) 'Cureghem en crise? Les processus de marginalisation d'un quartier bruxellois', in Liber Amicorum Herman Van der Haegen, *Acta Geographica Lovaniensia* 34: 361–74.
Mistiaen, P., Meert, H. and Kesteloot, C. (1995) 'Polarisation socio-spatiale et stratégies de survie dans deux quartiers bruxellois', *Espace–Populations–Sociétés* 3: 277–90.
Mollenkopf, J. and Castells, M. (1991) *Dual City: Restructuring New York*, New York: Russell Sage Foundation.
Montagné-Villette, S. (1990) *Le Sentier: un espace ambigu*, Paris: Mouton.
Morrill, R. (1995) 'Racial segregation and class in a liberal metropolis', *Geographical Analysis* 27: 22–41.
Morris, A. (1994) 'The desegregation of Hillbrow Johannesburg 1978–82', *Urban Studies* 31: 821–34.
Morris, P. (1980) *Soweto: A Review of Existing Conditions and Some Guidelines for Change*, Johannesburg: Urban Foundation.
Muller, Thomas (1993) *Immigrants and the American City*, New York: New York University Press.
Murdie, R. (1994a) 'Social polarisation and public housing in Canada: a case study of the Metropolitan Toronto Housing Authority', in F. Frisken (ed.) *The Changing Canadian Metropolis: A Public Policy Perspective*, vol. 1, Berkeley: Institute of Governmental Studies Press and Toronto: The Canadian Urban Institute, pp. 293–339.
— (1994b) '"Blacks in near-ghettos?" Black visible minority population in Metropolitan Toronto Housing Authority public housing units', *Housing Studies* 9: 435–57.
— (1996) 'Economic restructuring and social polarization in Toronto', in J. O'Loughlin and J. Friedrichs (eds) *Social Polarization in Post-Industrial Metropolises*, Berlin and New York: De Gruyter, ch. 9.
Murdie, R. and Borgegård, L.-E. (1992) 'Social differentiation in public rental housing: a case study of Swedish metropolitan areas', *Scandinavian Housing and Planning Research* 9: 1–17.
Murdie, R.A. and Northrup, D. (1989) *Residential Conversions in Toronto: The Availability of Rental Units in Owner Occupied Dwellings in the City of Toronto and Owners' Experience in the Rental Market*, Toronto: Institute for Social Research (York University).
Murie, A. (1983) *Housing Inequality and Deprivation*, London: Heinemann.
— (1993a) *Cities and Housing after the Welfare State*, Amsterdam: AME, University of Amsterdam.

—— (1993b) 'Restructuring housing markets and housing access', in R. Page and J. Baldock (eds) *Social Policy Review 5*, Social Policy Association.
Murie, A. and Musterd, S. (1996) 'Social segregation, housing tenure and social change in Dutch cities in the late 1980s', *Urban Studies* 3: 495–516.
Murtagh, B. (1994) *Public Sector Housing and Deprivation in Belfast*, Newtownabbey: University of Ulster Centre for Policy Research, Occasional Paper 6.
Musterd, S. and Ostendorf, W. (1993) 'Grote steden op de drempel tussen "veelkleurig" en "gepolariseerd"', *Geografie* 2, 3: 41–5.
—— (1994) 'Affluence, access to jobs and ethnicity in the Dutch welfare state: the case of Amsterdam', *Built Environment* 20, 3: 242–53.
—— (1995) 'Polarization and the Dutch welfare state: the case of Amsterdam', in G.O. Braun (ed.) *Managing and Marketing of Urban Development and Urban Life*, Berlin: Dietrich Reimer Verlag, pp. 315–28.
—— (1996) 'Amsterdam, urban change and the welfare state', in J. O'Loughlin and J. Friedrichs (eds) *Social Polarization in Post-Industrial Metropolises*, Berlin and New York: Walter De Gruyter, pp. 71–95.
Musterd, S., Ostendorf, W. and Breebaart, M. (1997) 'Segregation in European cities: patterns and policies', *Tijdschrift voor Economische en Sociale Geografie* 88, 2: 182–87.
Myrdal, Gunnar (1944) *An American Dilemma*, New York: Harper.
Naroska, Hans-Jürgen (1988) 'Urban Underclass und "neue" soziale Randgruppen im städtischen Raum', in Jürgen Friedrichs (ed.) *Soziologische Stadtforschung*, Opladen: Westdeutscher Verlag (Sonderheft 29 der Kölner Zeitschrift für Soziologie und Sozialpsychologie), pp. 251–71.
Nel, J. (1993) 'Ruimtelike segregasie en marginaliteit: die Chinese gemeenskap van Port Elizabeth', *South African Geographer* 20: 100–15.
Norcliffe, G., Goldrick, M. and Muszynski, L. (1986) 'Cyclical factors, technological change, capital mobility, and deindustralization in metropolitan Toronto', *Urban Geography* 7: 413–36.
Northeastern Illinois Planning Commission (1988) *NIPC Data Bulletin* 88–1, May, Chicago: Northeastern Illinois Planning Commission.
—— (1991) *NIPC Data Bulletin* 91–1, February, Chicago: Northeastern Illinois Planning Commission.
Northern Ireland Housing Executive (1972) *First Annual Report 1971–1972*, Belfast: Northern Ireland Housing Executive.
—— (1991) *Building a Better Belfast*, Belfast: Northern Ireland Housing Executive.
Northern Ireland Registrar General (1992) *Census of Population 1991: Belfast Urban Area Report*, Belfast: Her Majesty's Stationery Office.
O'Connor, J. (1989) 'Welfare expenditure and policy orientation in Canada in comparative perspective', *The Canadian Review of Sociology and Anthropology* 26: 127–50.
O'Connor, J. and Brym, R. (1988) 'Public welfare expenditure in OECD countries: towards a reconciliation of inconsistent findings', *The British Journal of Sociology* 39: 47–68.
Ogden, P. (1987) 'Immigration, cities and the geography of the National Front in France', in G. Glebe and J. O'Loughlin (eds) *Foreign Minorities in Continental European Cities*, Wiesbaden: Steiner Verlag, pp. 163–83.
Oliver, Q. (1992) 'The role of non-profit organisations in a divided society: the case of Northern Ireland', in K. McCarthy, K. Hodgkinson and V. Sumariwalla (eds) *The Non-Profit Sector in the Global Economy*, San Francisco: Jossey-Bass.
O'Loughlin, J. and Friedrichs, J. (1996) 'Polarisation in post-industrial societies: social and economic roots and consequences', in John O'Loughlin and Jürgen Friedrichs (eds) *Social Polarisation in Post-Industrial Metropolises*, Berlin and New York: Walter de Gruyter, pp. 1–18.
Olsson-Hort, S.-E. (1992) *Segregation ett Svenskt Dilemma? Bilaga 9 till LU 92*, Stockholm: Finansdepartementet.

Opoku-Dapaah, E. (1995) *Somali Refugees in Toronto: A Profile*, Toronto: York Lanes Press.

Oxley, H. and Martin, J. (1991) 'Controlling government spending and deficits: trends in the 1980s and prospects for the 1990s', *OECD Economic Studies* 17: 145–89.

Pahl, R. (1984) *Divisions of Labour*, Oxford: Basil Blackwell.

— (1988) 'Some remarks on informal work, social polarisation and the social structure', *IJURR* 12: 247–67.

Parkin, F. (1979) *Marxism and Class Theory: A Bourgeois Critique*, London: Tavistock.

Parnell, S. (1988) 'Land acquisition and the changing residential face of Johannesburg', *Area* 20: 307–14.

Peach, C. (1996) 'Does Britain have ghettoes?', *Transactions of the Institute of British Geographers* 12, 1: 214–35.

Penderis, S.P. and van der Merwe, I.J. (1994) 'Kaya Mandi hostels, Stellenbosch: place, people and policies', *South African Geographical Journal* 76: 33–8.

Pinch, S. (1993) 'Social polarization: a comparison of evidence from Britain and the United States', *Environment and Planning A* 25: 779–95.

Pirie, G.H. and Hart, D.M. (1985) 'The transformation of Johannesburg's black western areas', *Journal of Urban History* 11: 387–410.

Pirie, G.H. and da Silva, M. (1986) 'Hostels for African migrants in greater Johannesburg', *GeoJournal* 12: 173–80.

Platzky, L. and Walker, C. (1985) *The Surplus People: Forced Removals in South Africa*, Johannesburg: Ravan Press.

Polanyi, K. (1944) *The Great Transformation*, New York: Rinehart.

Power, A. (1996) 'Area-based poverty and resident empowerment', *Urban Studies* 33: 1535–64.

Power, J. and Shuttleworth, I. (1995) 'Segregation, de-industralisation and population loss: intercensal population change 1971–1991 in the Belfast Urban Area', paper read at International Conference on Population Geography, University of Dundee.

Preston, R. (1991) 'Central place theory and the Canadian urban system', in T. Bunting and P. Filion (eds) *Canadian Cities in Transition*, Toronto: Oxford University Press, ch. 7.

Preston, V., Murdie, R. and Northrup, D. (1993) 'Condominiums: an investment decision or lifestyle choice? A comparative study of resident and nonresident condominium owners in the city of Toronto', *Netherlands Journal of Housing and the Built Environment* 8: 281–300.

Preteceille, E. (1995) 'Division sociale de l'espace et globalisation: le cas de la metropole Parisienne', *Sociétés Contemporaines* 22/23, June/Sept.: 33–67.

Project Team (1977) *Belfast: Areas of Special Social Need*, Belfast: Her Majesty's Stationery Office.

Ram, B., Norris, M.J. and Skof, K.J. (1989) *The Inner City in Transition*, Ottawa: Statistics Canada.

Ray, B. (1994) 'Immigrant settlement and housing in metropolitan Toronto', *The Canadian Geographer* 38: 262–5.

Reardon, P. (1993) 'More Chicagoans find it isn't their kind of town', *The Chicago Tribune*, 28 November 1993.

Reddin, M. (1970) 'Some relationships between income taxation and social security', *International Social Security Review* 1: 113–20.

Reed, A., Jr (1988) 'The black urban regime: structural origins and constraints', in M.P. Smith (ed.) *Power, Community, and the City. Comparative Urban and Community Research*, vol. 1, New Brunswick, NJ: Transaction.

Regionplane- och trafikkontoret (1995) *Storstockholms Sociala Geografi*, Stockholm: Regionplane- och trafikkontoret.

Reitz, J. (1990) 'Ethnic concentrations in labour markets and their implications for ethnic inequality', in R. Breton, W.W. Isajiw, W.E. Kalbach and J.G. Reitz (eds)

Ethnic Identity and Equality: Varieties of Experience in a Canadian City, Toronto: University of Toronto Press, ch. 4.
Rhein, C. (1996) 'Social segmentation and spatial polarization in Greater Paris', in John O'Loughlin and Jürgen Friedrichs (eds) *Social Polarisation in Post-Industrial Metropolises*, Berlin and New York: Walter de Gruyter, pp. 45–70.
Richards, J. (1994) 'The social policy round', in W.G. Watson, J. Richards and D.M. Brown (eds) *The Case for Change: Reinventing the Welfare State*, Toronto: C.D. Howe Institute.
Richmond, A. (1992) 'Immigration and structural change: the Canadian experience, 1971–1986', *International Migration Review* 26: 1200–21.
Ricketts, E. and Sawhill, I. (1988) 'Defining and measuring the underclass', *Journal of Policy Analysis and Management* 7: 313.
Ringqvist, M. (1996) *Om den Offentliga Sektorn. Vad den Ger och vad den Tar*, Stockholm: Publica–Fritzes.
Riordan, R. (1988) 'The Ukuhleleleka of Port Elizabeth', *Monitor* 1: 2–16.
Robson, B. (1988) *Those Inner Cities*, Oxford: Clarendon Press.
Robson, B., Bradford, M. and Deas, I. (1994) *Relative Deprivation in Northern Ireland*, Belfast: Policy Planning and Research Unit.
Roof, Wade C., Van Valey, Thomas L. and Spain, Daphne (1976) 'Residential segregation in southern cities: 1970', *Social Forces* 55: 59–71.
Rusk, D. (1993) *Cities without Suburbs*, Washington: The Woodrow Wilson Center Press.
Sahlin, I. (1996) *På Gränsen till Bostad. Avvisning, Utvisning, Specialkontrakt*, Lund: Arkiv Förlag.
Sassen, Saskia (1988) *The Mobility of Labor and Capital*, Cambridge: Cambridge University Press.
—— (1991) *The Global City, New York, London, Tokyo*, Princeton, NJ: Princeton University Press.
—— (1994a) *Cities in World Economy*, Thousand Oaks: Pine Forge Press.
—— (1994b) 'Ethnicity and space in the global city: a new frontier', in Seamus Dunn (ed.) *Managing Divided Cities*, Keele, Staffordshire: Ryburn Publishing/Keele University Press, pp. 13–29.
—— (forthcoming) 'Identity in the global city: economic and cultural encasements', in Anthony King (ed.) *Re-Presenting the City*, London: Macmillan.
Sassen-Koob, S. (1984) 'The new labor demand in global cities', in M.P. Smith (ed.) *Cities in Transformation*, Beverly Hills: Sage, Urban Affairs Annual 26, pp. 139–71.
Schill, M. (1993) 'Distressed public housing: where do we go from here?', *The University of Chicago Law Review* 60: 497–554.
—— (1994) 'Race, the underclass, and public policy', *Law and Social Inquiry* 19, 2: 433–56.
Schüler, Horst (1994) 'Einkommen und seine Verteilung in Hamburg 1989', *Hamburg in Zahlen* 4: 112–19.
Scott, J. (1994) *Poverty and Wealth: Citizenship, Deprivation and Privilege*, London/New York: Longman Sociology.
See, K. O'Sullivan (1991) 'Comments from the special issue editor: approaching poverty in the United States', *Social Problems* 38: 427–31.
Sellers, Jeffrey M. (1994) 'Norms in the politics of housing for minorities in a French, a German and an American metropolitan area', paper presented at the annual meeting of the American Political Science Association, New York.
Semple, R.K. (1988) 'Urban dominance, foreign ownership, and corporate concentration', in J. Curtis, E. Grabb, N. Gruppy and S. Gilbert (eds) *Social Inequality in Canada: Patterns, Problems and Policies*, Scarborough, Ontario: Prentice-Hall Canada, ch. 29.

Shevky, E. and Bell, W. (1955) *Social Area Analysis*, Stanford: Stanford University Press.
Short, J. (1989) 'Yuppies, yuffies and the new urban order', *Transactions of the Institute of British Geographers* 14: 173–88.
Silver, H. (1993) 'National conceptions of the new urban poverty: social structural change in Britain, France and the United States', *International Journal of Urban and Regional Research* 17, 3: 336–54.
Simon, D. (1989) 'Crisis and change in South Africa: implications for the apartheid city', *Transactions of the Institute of British Geographers* (New Series) 14: 189–206.
Smith, A.D. (1986) *The Ethnic Origins of Nations*, Oxford: Blackwell.
Smith, D.M. (1992) *The Apartheid City and Beyond: Urbanization and Social Change in South Africa*, London: Routledge.
Sociaal en Cultureel Planbureau (1992) *Sociaal en Cultureel Rapport 1992*, Den Haag/Rijswijk: VUGA.
—— (1996) *Sociaal en Cultureel Rapport 1996*, Rijswijk: Sociaal en Cultureel Planbureau.
Social rapport (1994) *SoS rapport 1994*: 10, Stockholm: Socialstyrelsen.
Soni, D. and Maharaj, B. (1991) 'Emerging urban forms in rural South Africa', *Antipode* 23: 47–67.
SOU (1974) *Orter i Regional Samverkan. Separatkartor. Betänkande av Expertgruppen för Regional Utredningsverksamhet. SOU 1974: 10 Appendix*, Stockholm: Allmänna Förlaget.
—— (1990a) *Storstadsliv. Rika Möjligheter – Hårda Villkor. Slutbetänkande från Storstadsutredningen. 1990: 36*, Stockholm: Statsrådsberedningen, Fritzes.
—— (1990b) *Välfärd och Segregation i Storstadsregionerna; Underlagsrapport. Storstadsutredningen 1990:20,* Stockholm: Fritzes.
—— (1993) *Socialtjänstens Roll i Samhällsplanering och Samhällsarbete: en Kunskapsöversikt och ett Diskussionsunderlag: Rapport 1993: 91,* Stockholm: Allmänna Förlaget.
—— (1996: *Bostadspolitik 2000 – från Produktions- till Boendepolitik. Slutbetänkande av Bostadspolitiska Utredningen*, Stockholm: Fritzes.
South, Scott J. and Deane, Glenn D. (1993) 'Race and residential mobility: individual determinants and structural constraints', *Social Forces* 72: 147–68.
South Africa (1992a) *Population Census 1991: Adjustment for Undercount*, Report 03–01–26, Pretoria: Government Printer.
—— (1992b) *Population Census 1991: Selected Statistical Region-Port Elizabeth-Uitenhage*, Report 03–01–13, Pretoria: Government Printer.
Sporton, D. and White, P. (1989) 'Immigrants in social housing: integration or segregation in France?', *The Planner* 75, 4: 28–31.
Squires, Gregory (1994) *Capital and Communities in Black and White*, Albany: SUNY Press.
Stanback, T.M., Jr and Noyelle, T.J. (1982) *Cities in Transition: Changing Job Structures in Atlanta, Denver, Buffalo, Phoenix, Columbus (Ohio), Nashville*, Charlotte, NJ: Allenheld, Osmun.
Stark, T. (1992) *Income and Wealth in the 1980s*, 3rd edn, London: Fabian Society.
Statistics Sweden (1995) *Inkomstfördelningsundersökningen 1993*, Örebro: Statistics Sweden, Publication Service.
—— (1996) *Statistical Yearbook of Sweden 1996*, Örebro: Statistics Sweden, Publication Service.
Stockholm Office of Integration (1996) *Stockholm- Bilder av Segregation*, Stockholm: Stockholm Office of Integration.
Stovall, T. (1990) *The Rise of the Paris Red Belt*, Berkeley: University of California Press.
Sweeney, P. and Gaffikin, F. (1995) *Listening to People*, Belfast: Making Belfast Work.
Swilling, M., Humphries, R. and Shubane, K. (1991) *Apartheid City in Transition*, Cape Town: Oxford University Press.

Taeuber, Karl E. and Taeuber, Alma F. (1965) *Negroes in Cities*, New York: Aldine.
Teixeira, C. (1995) 'The Portuguese in Toronto: a community on the move', *Portuguese Studies Review* 4: 57–75.
Tigges, Leann M. and Tootle, Deborah M. (1993) 'Underemployment and racial competition in local labor markets', *Sociological Quarterly* 34: 279–98.
Timms, D.W.G. (1971) *The Urban Mosaic*, Cambridge: Cambridge University Press.
United Nations Development Programme (1994) *Human Development Report, 1994*, New York: Oxford University Press.
Urban Studies (1994) 31.
US Bureau of the Census (1992) *Statistical Abstract of the United States, 1992*, Washington, DC: US Government Printing Office.
Van Kempen, R. (1994) 'Ruimtelijke segregatie, ruimtelijke concentratie en sociale marginalisering in de Nederlandse stad', *Planologisch Nieuws* 14, 4: 367–77.
Vandermotten, C. (1994) 'Le plan régional de développement de la région de Bruxelles-capitale', in C. Vandermotten (ed.) *Planification et stratégies de développement dans les capitales européennes*, Brussels: Edition de l'Université de Bruxelles, pp. 195–206.
Vant, A. (1986) *Marginalité sociale: marginalité spatiale*, Paris: Editions du CNRS.
Veser, Jürgen (1991) 'Das Abschmelzen des Sozialwohnungsbestandes. Ursachen und Auswirkungen auf unterschiedlichen regionalen Wohnungsmärkten', *Informationen zur Raumentwicklung* 5/6: 359–78.
Vieillard-Baron, H. (1992) 'Deux ZAC de banlieue en situation extrême: du grand ensemble stigmatisé de Chanteloup au "village" de Chevry', *Annales de Géographie* 101: 188–213.
Vogel, J., Andersson, L.-G., Davidsson, U. and Häll, L. (1987) *Living Conditions. Report no 51. Inequality in Sweden 1975–85*, Stockholm: Statistics Sweden.
Vos, S. de (1997) *De Omgeving Telt. Compositionele Effecten in de Sociale Geografie*, Amsterdam: Thesis.
Vranken, J. and Geldof, D. (1992) *Armoede en Sociale Uitsluiting, Jaarboek 1991*, Leuven/Amersfoort: Acco.
—— (1993) *Armoede en Sociale Uitsluiting, Jaarboek 1992–1993*, Leuven/Amersfoort: Acco.
Vranken, J., Geldof, D. and Van Menxel, G. (1994) *Armoede en Sociale Uitsluiting, Jaarboek 1994*, Leuven/Amersfoort: Acco.
—— (1995) *Armoede en Sociale Uitsluiting, Jaarboek 1995*, Leuven/Amersfoort: Acco.
—— (1996) *Armoede en Sociale Uitsluiting, Jaarboek 1996*, Leuven/Amersfoort: Acco.
Wacquant, L. (1989) 'The ghetto, the state, and the new capitalist economy', *Dissent* Fall: 508–12.
—— (1993) 'Urban outcasts: stigma and division in the Black American ghetto and the French urban periphery', *International Journal of Urban and Regional Research* 17, 3: 366–83.
—— (1994) 'The hyperghetto as structure and practice', lecture presented at the conference on New Conceptions of Urban Space, Columbia University Graduate School of Architecture, Planning and Preservation, 14 October.
Wacquant, L. and Wilson, W. (1989) 'Poverty, joblessness, and the social transformation of the inner city', in P. Cottingham and D. Ellwood (eds) *Welfare Policy for the 1990s*, Cambridge, MA: Harvard University Press.
—— (1993) 'The cost of racial and class exclusion in the inner city', in W.J. Wilson (ed.) *The Ghetto Underclass: Social Science Perspectives*, Newbury Park: Sage, pp. 25–42.
Waldorf, Brigitte S. (1993) 'Segregation in urban space: a new measurement approach', *Urban Studies* 30: 1151–64.
Watson, W.G. (1994) 'The view from the right', in W.G. Watson, J. Richards and D.M. Brown (eds) *The Case for Change: Reinventing the Welfare State*, Toronto: C.D. Howe Institute.

Weber, M. (1968) *Economy and Society* (eds) G. Roth and C. Wittick, New York: Bedminster Press.

Western, J. (1981) *Outcast Cape Town*, Minneapolis: University of Minnesota Press.

White, P.E. (1991) 'The dwellings of the poor in Paris, 1789–1989', in D. Williams (ed.) *1789, the Long and the Short of It*, Sheffield: Sheffield Academic Press, pp. 139–58.

White, P.E. and Winchester, H.P.M. (1991) 'The poor in the inner city: stability and change in two Parisian neighbourhoods', *Urban Geography* 12: 35–54.

Wiley, Norbert F. (1970) 'The ethnic mobility trap and stratification theory', in Peter I. Rose (ed.) *The Study of Society*, New York: Random House, pp. 397–408.

William-Olsson, W. (1961) *Stockholm: Its Structure and Development,* Uppsala: Almqvist & Wiksell.

Williams, N. and Twine, F. (1990) *The Resale of Council Dwellings*, Scottish Homes.

Wilson, W.J. (1978) *The Declining Significance of Race: Blacks and Changing American Institutions*, Chicago: University of Chicago Press.

— (1987) *The Truly Disadvantaged: The Inner City, the Underclass, and Public Policy*, Chicago: University of Chicago Press.

— (1990) 'The underclass: issues, perspectives and public policy', *Annals* 501: 26–47.

— (1994) 'Crisis and challenge: race and the new urban poverty', the 1994 Ryerson lecture, University of Chicago, 8 April 1994.

Winchester, H.P.M. and White, P.E. (1988) 'The location of marginalised groups in the inner city', *Environment and Planning D: Society and Space* 6: 37–54.

Winnick, Louis (1990) *New People in Old Neighborhoods*, New York: Russell Sage.

Work and Leisure (1993) *The National Atlas of Sweden, Work and Leisure*, Stockholm: Almqvist & Wiksell.

Wright, E. and Martin, B. (1987) 'The transformation of the American class structure, 1960–1980', *American Journal of Sociology* 93: 1–29.

Yeates, M. (1991) 'The Windsor–Quebec corridor', in T. Bunting and P. Filion (eds) *Canadian Cities in Transition*, Toronto: Oxford University Press, ch. 8

INDEX